Green Energy and Technology

Climate change, environmental impact and the limited natural resources urge scientific research and novel technical solutions. The monograph series Green Energy and Technology serves as a publishing platform for scientific and technological approaches to "green"—i.e. environmentally friendly and sustainable—technologies. While a focus lies on energy and power supply, it also covers "green" solutions in industrial engineering and engineering design. Green Energy and Technology addresses researchers, advanced students, technical consultants as well as decision makers in industries and politics. Hence, the level of presentation spans from instructional to highly technical.

Indexed in Scopus.

Indexed in Ei Compendex.

Jiefeng Hu
Editor

Community Energy and Microgrids

Control, Operation and Optimization

Springer

Editor
Jiefeng Hu
Centre for New Energy Transition Research
Federation University Australia
Ballarat, VIC, Australia

ISSN 1865-3529 ISSN 1865-3537 (electronic)
Green Energy and Technology
ISBN 978-981-97-6296-5 ISBN 978-981-97-6297-2 (eBook)
https://doi.org/10.1007/978-981-97-6297-2

This Springer imprint is published by the registered company Springer Nature Singapore Pte Ltd.
The registered company address is: 152 Beach Road, #21-01/04 Gateway East, Singapore 189721, Singapore

Preface

Increasing penetration of distributed energy resources (DERs) in the electrical power network has revolutionized the way that electricity is generated, transferred, and consumed. In this context, microgrids are being developed and expanded as an essential and fundamental building block for future smart power grids. Specifically, community microgrids are emerging as a solution for neighborhoods, villages, towns, and cities to meet their energy needs locally. As their name implies, they serve entire communities. A community microgrid is a local power grid covering one or more distribution substations and supported by high penetrations of local DERs, including renewable energy sources. It provides cost-effective electricity services to consumers and prosumers and increases energy self-reliance.

This book offers readers opportunities to explore community energy and microgrids with conceptual illustrations, investigations, technical aspects, and examples in the changing energy landscape. It covers a comprehensive study of the control, operation, and optimization of community energy and microgrids, ranging from distributed generations, power electronic converters, power quality, load forecasting, fault protection, and communication, to the optimization of operation. This book is divided into 8 chapters. Chapter 1 introduces the fundamentals of DERs and the evolution of conventional power grids toward modern smart grids with high penetrations of DERs. Chapter 2 describes the key features of community microgrids as well as their electrical configurations, and then highlights the challenges in operations and discusses the recent development of community microgrids in the energy sector. Chapter 3 describes the role of grid-forming and grid-following inverters, and then introduces a master–slave model predictive control strategy to coordinate distributed inverters in microgrids. Chapter 4 reviews the three-layer hierarchical control framework applied to microgrids. The shortcomings of the primary control are revealed. After that, a distributed secondary control scheme with inertia emulation is presented to improve frequency regulation. Chapter 5 reveals the challenges and power quality issues of grid-connected inverter operation under weak power grids. Subsequently, the capacitor voltage Lead compensation feedforward method and the PCC voltage control reactive power compensation method are used to solve the LCL resonance problem and the PCC voltage drop problem when the three-phase grid-connected inverter operates under a weak power grid. Also, a grid impedance detection method based on a complex coefficient filter is developed. In Chap. 6, various load forecasting technologies are reviewed. After that, a load probabilistic forecasting technique based on a neural network is introduced for microgrid applications. Chapter 7 highlights the importance of communication networks in microgrids for monitoring and control. Then, a distributed bias min-consensus (DBMC)-based approach is proposed to optimize the WAMS communication network. The objective is to minimize the ETE delay and maximize the data arrival rate received by the PDC under normal conditions and cyberattacks. Chapter 8 presents an optimization scheme to manage distributed energy resources of a community-scale multi-energy system to both customers and network operators.

Melbourne, Australia
October 2023

Jiefeng Hu

Contents

Chapter 1
Introduction

Jiefeng Hu, Josep Guerrero, and Jose Rodriguez

1.1 Distributed Energy Resources

The electrical power system is currently under major transformation from electricity generation, and distribution, to consumption. This is mainly driven by the net zero emission target, the rapid depletion of conventional fossil fuels, as well as the increasing electricity all around the world. Over the last decades, distributed energy resources (DERs) including renewable and non-renewable sources have been exploited rapidly, such as solar photovoltaic (PV) systems, gas microturbine generators, wind generators, gas/steam-powered combined heat and power (CHP) stations, fuel cells, etc. These local DERs offer alternative and economic power generation options with high efficiency and reduced losses [1].

DER is an electric power generation source that is connected to the distribution network or just to a local load like the uninterruptible power supply (UPS) systems. The advantages of DERs include but not limited to:

- High reliability of power supply, especially when interruption of service is unacceptable.
- An easy way to integrate renewable energy sources
- Flexible in deployment with plug-and-play capability

J. Hu (✉)
Centre for New Energy Transition Research, Federation University Australia, Mount Helen, VIC 3353, Australia
e-mail: j.hu@federation.edu.au

J. Guerrero
Technical University of Catalonia, Barcelona, Spain & Aalborg University, Aalborg East, Denmark
e-mail: joz@energy.aau.dk

J. Rodriguez
Universidad San Sebastian, Santiago, Chile
e-mail: jose.rodriguezp@uss.cl

J. Hu (ed.), *Community Energy and Microgrids*, Green Energy and Technology,
https://doi.org/10.1007/978-981-97-6297-2_1

- Effective electricity supply for remote areas without utility grid infrastructures.

In practice, DERs can be offered in different types and forms. Some of the most promising and popular ones for microgrid applications are highlighted as follows.

Wind

Wind energy can be captured by a horizontal axis or vertical axis wind turbine, and then converted into electricity through an electrical generator [2, 3]. A typical modern wind power system usually consists of mechanical power conversion (e.g., rotor, blade, yaw system, and gearbox), electrical power conversion (e.g., generator, power electronic interface, and transformer) as well as monitoring, control, and communication. With the transition to net zero, there will be an increase in demand for renewable energies. As wind energy is sustainable and clean, it has become one of the fastest-growing and most promising energy resources in the world. Meanwhile, because of the maturity of wind technologies and related governmental policies, the cost of electricity generation from wind energy has been reduced gradually. In 2022, 77.6 GW of additional wind capacity was added to power grids, lifting total installed wind capacity up to 906 GW, a rise of 9% compared with 2021 [4]. In 2023, with the further ease of COVID-19 restrictions, onshore wind capacity additions were able to rebound quickly, which claims the second highest growth among all renewable power technologies, just behind solar PVs [5]. Forward-looking into 2030, GWEC Market Intelligence expects an additional 143 GW by the end of the decade [4].

Solar

Solar energy can be harvested in various ways and then converted into different forms of energy. One of the most popular means is the conversion of energy from sunlight into electricity using solar PVs. In 2022, solar PV generation grew by a record 270 TWh, totaling almost 1 300 TWh. This marks a significant milestone for solar PV generation as it surpassed wind for the first time in history to become the largest absolute generation growth among all renewable technologies. More excitingly, it is predicted that solar PV's installed power capacity will likely overtake that of natural gas by 2026 and coal by 2027, becoming the largest of all [5]. From a technical viewpoint, solar PV technology essentially converts solar energy directly into electrical energy by means of solar cells, which are made of semiconductor materials such as silicon. To increase the power output, these solar cells are usually combined into modules with unique voltage and current characteristics. Further, the modules can be grouped together in various configurations and quantities to form arrays with larger output voltages and currents [6, 7]. Solar panels can be manufactured in a very modular way, which provides flexibility in planning and financial arrangement. As a result, they can be deployed in a wide range of applications, from residential rooftops with small quantities to utility-scale solar farms with large capacities.

Combustion Turbines

While much of the recent research attention naturally focuses on green and emerging technologies such as solar and wind power generation, the reality is that conventional combustion turbines will continue to play an important role in the rollout of distributed generations as well as microgrids. Small size and capacity combustion turbines, which consume fossil fuels such as propane, natural gas, petrol, etc., can be flexibly employed for distributed power generation in microgrids [8]. For instance, a gas microturbine consists of rotating blades, a compressor, a combustion chamber, and certainly an electric machine. These components work together to convert fossil energy to mechanical energy and then to electric energy. These microturbines can serve as backup power supply or peak load support.

Hydrogen

When methane, also known as natural gas, is burnt, it provides high-temperature and high-pressure steam to spin a turbine that drives a generator to produce electricity. However, gas turbine power plants release a large amount of waste product, i.e., carbon dioxide. Hydrogen, on the other hand, is a clean alternative to methane as burning hydrogen does not lead to carbon dioxide emissions. With the net zero emission target in place, there is a potential for large-scale hydrogen applications. In terms of hydrogen utilization, proton exchange membrane fuel cells (PEMFC) are competitive means, by which hydrogen reacts with oxygen across an electrochemical cell, producing small amounts of heat, water, and most importantly, electricity [9]. To ensure a sufficient and sustainable hydrogen supply chain, the production, storage, and transportation of hydrogen becomes critically important. Although hydrogen is a clean alternative to generating electricity by using fuel cells, most of the world's current hydrogen supply is still obtained by steam-reforming fossil fuels. By contrast, green hydrogen in the industry can be obtained by electrolysis from green electricity. Recently, increasing attention has been paid to renewable energy sources, in which wind and solar are coupled with fuel cells to provide fully clean and sustainable electric power.

Energy Storage

With more and more intermittent renewables and inertialess power electronic converters connected to the power network, energy storage has become increasingly crucial in microgrids. Energy storage can help increase renewable penetration, improve energy efficiency, and enhance grid stability as well as resilience. Essentially, it provides a variety of technological solutions to fill the gap between power supply and load demand to enable a resilient power infrastructure that can be beneficial to both consumers and utilities [10]. From a more technical point of view, energy storage, together with power electronic equipment, plays a critical role in stabilizing the voltage and frequency of the microgrid by active-reactive power management, particularly in islanded mode. It serves as an energy buffer to alleviate load uncertainties and compensate renewable intermittency and hence, improve microgrid stability.

Energy storage technologies are currently being developed at different levels of technological maturity, including mechanical storage, chemical storage, thermal storage, electrochemical storage, etc. Among them, some are already proven for commercial-scale applications such as batteries and pumped hydropower. Nowadays, batteries are the most popular and widely used energy storage systems in microgrid applications [11]. To cope with the intermittent nature of renewable energy, batteries can be used to absorb surplus energy during peak generation periods while providing additional power in demand peak.

1.2 Microgrids and Smart Grids

A traditional power grid essentially consists of three major components:

- central power plants (hydro, coal, gas, and nuclear power stations);
- electricity delivery infrastructure (towers, poles, and wires); and
- energy consumption (homes, manufacturers, businesses, etc.).

In recent decades, however, a fourth element has crept into traditional power grids: small-to-medium-scale DERs. Besides, increasing electricity demand, and depletion of fossil fuels, together with net zero policies, have driven the rapid deployment of renewable power generations. Rooftop solar PV systems, for example, are the most common small-scale residential distributed generation. The rise of these affordable small-scale renewable energy is changing energy systems globally. The existing electrical power network is transforming from a conventional centuries-long centralized structure toward a decentralized architecture with increasing levels of DERs. Nowadays, it is possible for households and commercial buildings to affordably meet a portion of energy needs themselves rather than relying solely on utilities.

To facilitate the integration of the DERs discussed above, microgrids have emerged as an effective solution. According to the definition from the US Department of Energy, a microgrid is a cluster of interconnected DERs and loads within clearly defined electrical boundaries that acts as a single controllable entity [12]. As the building blocks of next-generation smart grids, a microgrid is a self-contained power grid that uses a variety of DERs to generate electric power. It can operate either in islanded or grid-connected modes. It offers distinct merits over conventional power grids in such a way that distributed non-renewable and renewable energy sources can be flexibly integrated and deployed [13, 14]. Consequently, microgrids can help reduce carbon emissions, improve the reliability and resilience of the electricity supply, and hence improve sustainability. They can complement the national power grid or work independently from it, providing communities with access to more sustainable and cost-effective electricity supplies.

In general, three major objectives should be achieved in order to ensure the proper operation of a microgrid, namely: (1) load sharing among DERs proportionately according to their capacities [15, 16], (2) stable frequency and voltage supply at the point of common coupling [17–19], and (3) operation among DERs as well as between microgrid and utility for overall benefit maximization [20]. Usually, the first two objectives can be fulfilled by controlling power electronic converters locally to regulate the output currents and voltages. By contrast, the third task aims to coordinate energy dispatch to maximize renewable utilization, mitigate power loss, reduce operational costs, etc. while satisfying various constraints. With the latest technology innovations, such as 5G/6G connectivity, internet of Things (IoT) systems [21], and artificial intelligence (AI) [22–24], we can connect, monitor and manage renewable sources and ensure a more efficient, reliable and sustainable microgrid, and subsequently, a large-scale smart grid infrastructure with interconnected microgrids.

References

1. J. Hu, Y. Shan, K. Cheng, S. Islam, Overview of power converter control in microgrids—challenges, advances, and future trends. IEEE Trans. Power Electron. **37**(8), 9907–9922 (2022)
2. J. Hu et al., Simple and robust predictive direct control of DFIG with low constant switching frequency and reduced torque and flux ripples, in Proceedings of IEEE International Electric Machines & Drives Conference (IEMDC), pp. 771–776 (2011)
3. J. Hu et al., A novel control strategy for doubly fed induction generator and permanent magnet synchronous generator during voltage dips, in Proceedings of 20th Australian Universities Power Engineering Conference, pp. 1–6 (2010)
4. https://www.iea.org/energy-system/renewables/wind
5. Global Wind Energy Council, Global Wind Report (2023)
6. J. Hu, Y. Shan, Y. Xu, J.M. Guerrero, A coordinated control of hybrid ac/dc microgrids with PV-wind-battery under variable generation and load conditions. Int. J. Electr. Power Energy Syst. **104**, 583–592 (2019)
7. J. Hu, Y. Xu, K.W. Cheng, J.M. Guerrero, A model predictive control strategy of pv-battery microgrid under variable power generations and load conditions. Appl. Energy **221**, 195–203 (2018)
8. E. Konečná, S.Y. Teng, V. Máša, New insights into the potential of the gas microturbine in microgrids and industrial applications. Renew. Sustain. Energy Rev. **134** (2020)
9. G.M. Masters, *Renewable and efficient electric power systems*. IEEE Wiley, ISBN: 9781118140628 (2013)
10. CSIRO, Renewable Energy Storage Roadmap (2023)
11. M. T. Lawder et al., Battery energy storage systems (BESS) and battery management system (BMS) for grid-scale applications. IEEE Proc. **102**(6) (2014)
12. https://www.energy.gov/oe/activities/technology-development/grid-modernization-and-smart-grid/role-microgrids-helping
13. J. Hu, Y. Shan, J.M. Guerrero, A. Ioinovici, K.W. Chan, J. Rodriguez, Model predictive control of microgrids—an overview. Renew. and Sustain. Energy Rev. **136**(5), 1–12 (2021)
14. J. Hu, J.M. Guerrero, S. Islam, Model Predictive Control for Microgrids: From power electronic converters to energy management, UK: IET (2021)
15. Y. Shan, J. Hu, M. Liu, J. Zhu, J.M. Guerrero, Model predictive voltage and power control of Islanded PV-battery microgrids with washout filter based power sharing strategy. IEEE Trans. Power Electron. **35**(2), 1227–1238 (2020)
16. J. Hu, et al., A new virtual-flux-vector based droop control strategy for parallel connected inverters in microgrids, in Proceedings of IEEE ECCE Asia Downunder, pp. 585–590 (2013)
17. Y. Shan, J. Hu, J.M. Guerrero, A model predictive power control method for pv and energy storage systems with voltage support capability. IEEE Trans. Smart Grid **11**(2), 1018–1029 (2020)

18. Y. Shan, J. Hu, K.W. Chan, Q. Fu, J.M. Guerrero, Model predictive control of bidirectional DC–DC converters and AC/DC interlinking converters—a new control method for PV-Wind-Battery microgrids. IEEE Trans. Sustain. Energy **10**(4), 1823–1833 (2019)
19. Y. Shan, J. Hu, Z. Li, J.M. Guerrero, A model predictive control for renewable energy based ac microgrids without any PID regulators. IEEE Trans. Power Electron. **33**(11), 9122–9126 (2018)
20. J. Hu et al., Economic model predictive control for microgrid optimization: a Review. IEEE Trans. Smart Grid **15**(1), 472–484 (2024)
21. Y. Saleem et al., Internet of Things-aided smart grid: technologies, architectures, applications, prototypes, and future research directions. IEEE Access. **7**, 62962–63003 (2019)
22. Y. Shan, J. Hu, B. Shen, Distributed secondary frequency control for ac microgrids using load power forecasting based on artificial neural network. IEEE Trans. Ind. Inform., in press
23. Y. Shan, J. Hu, H. Liu, A holistic power management strategy of microgrids based on Model Predictive Control and particle swarm optimization. IEEE Trans. Ind. Inform. **18**(8), 5115–5126 (2022)
24. Y. Shan, J. Hu, K. Chan, S. Islam, A unified model predictive voltage and current control for microgrids with distributed fuzzy cooperative secondary control. IEEE Trans. Ind. Inform. **17**(12), 8024–8034 (2021)

Chapter 2
Community Microgrids

Jiefeng Hu, Syed Islam, Iven Mareels, and Jianguo Zhu

2.1 Features and Configurations of Community Microgrids

According to the scale of the power capacity and the size of the service provided, microgrids can generally be categorized into different types, e.g., solar homes, campus and institutional microgrids, remote area microgrids, community microgrids, commercial and industrial microgrids, and military base microgrids [1]. As the name indicates, community microgrids are designed and constructed for a designated community. Usually, rural and regional communities tend to be at the end of the grid or have long distribution lines, leading to quality deterioration of the electricity supply. To address this problem, community microgrids provide reliable electricity services to households, villages, and towns locally. They can be further interconnected for financial and/or geographical reasons, e.g., to reduce costs by exploiting time-of-use electricity pricing. Community microgrids can even be deployed in regions with a satisfactory electricity supply from the main grid, mainly as an alternative to enhance local energy resilience and independence [2]. Community microgrids may or may not have a connection point to the main grid. The aim is to utilize local small-to-medium-scale distributed energy resources (DERs) to increase electrical energy self-reliance.

J. Hu (✉) · S. Islam · I. Mareels
Federation University Australia, Mount Helen, VIC 3353, Australia
e-mail: j.hu@federation.edu.au

S. Islam
e-mail: s.islam@federation.edu.au

I. Mareels
e-mail: i.mareels@federation.edu.au

J. Zhu
The University of Sydney, Camperdown, NSW 2050, Australia
e-mail: jianguo.zhu@sydney.edu.au

J. Hu (ed.), *Community Energy and Microgrids*, Green Energy and Technology,
https://doi.org/10.1007/978-981-97-6297-2_2

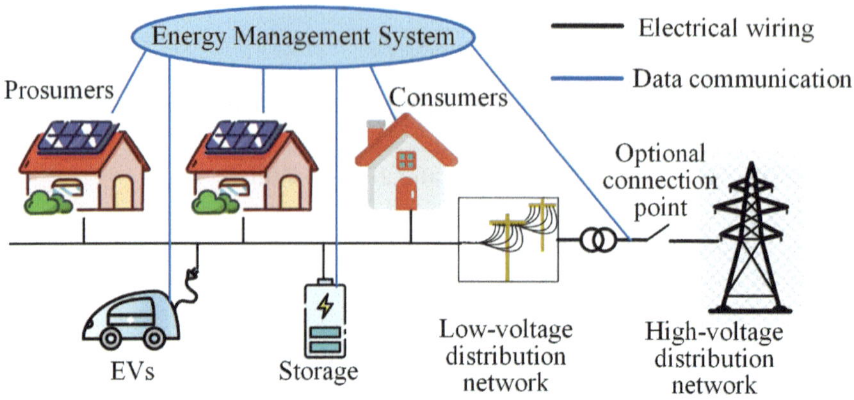

Fig. 2.1 Schematic diagram of a community microgrid

From an electrical perspective, a community microgrid is a localized power network served by one or more distribution substations integrated with high penetrations of local renewable and non-renewable DERs, energy storage, and energy management systems to provide cost-effective electricity services to consumers and prosumers, as illustrated in Fig. 2.1.

The basic components of a community microgrid include:

- **Local Generation**: Any type of generator may have a place on a community microgrid, including various DERs, as described previously in Chapter One. For example, solar PVs, wind generators, waste-to-energy, hydrogen fuel cells, biomass, and small-scale hydropower plants may feed into a community microgrid. In addition, diesel generators and gas turbines may be used as backups to increase resilience further. It is worth pointing out that under a net zero emission policy, renewable energy sources become increasingly appealing.
- **Consumption**: Houses, buildings, centers, farms, and transportation that use electricity can be considered loads in a community microgrid. Some may also be considered in the demand–supply energy balance, in particular, when the service is more related to energy rather than power (e.g., the fridge, where cooling energy is more important than cooling power). With the rapid acceleration of electrification, many new modern electric appliances have been seen in different areas, from consumer electronics, manufacturing equipment, and transportation to life support systems.
- **Energy Storage**: As emphasized in Chapter One, energy storage can help increase renewable penetration, improve energy efficiency, and enhance grid stability and resilience. With increasingly affordable but intermittent renewable energy generators connected to the community microgrids, energy storage plays a more significant role, and all forms of storage may be considered mechanical (like flywheels), chemical (e.g., hydrogen), thermal (e.g., ground source heat pump), or electrochemical (e.g., battery) storage. Based on the present market, batteries are one of

the most affordable and versatile energy storage means in microgrids. A battery bank can store excess electrical energy in times of surplus and rerelease it during peak demand. Furthermore, with an increasing number of electric vehicles (EVs) continuing to reshape the transportation sector and the distribution network in the coming decades, EVs are expected to make more contributions to demand-side management through optimal time-shift charging/discharging as portable energy storage [3, 4].

- **Distribution Network**: From the viewpoint of energy delivery, a microgrid is an electric network that forms physical power paths connecting power sources and loads. The existing electrical distribution network can be adopted in community microgrid planning and construction. Nevertheless, it may need to be upgraded to include additional equipment, e.g., smart sensors, circuit breakers, relays, contactors, etc., to facilitate plug-and-play operations of DERs and grid connection/disconnection of the entire microgrid.
- **Energy Management System (EMS)**: Using advanced control and communications technologies, a community microgrid will need a sensor-communication-decision architecture to pool all information together and then make decisions. EMS, which takes the role of decision-making, either from a central controller or distributed controllers, helps to use DERs and manipulate loads as cost-effectively and efficiently as possible. It collects real-time information from on-site sensors to decide the optimized operation and then sends the control signals back to each component. It coordinates supply and demand-side management, subject to certain system constraints, to achieve a safe, economical, and reliable microgrid operation. It aims to achieve various objectives, from energy savings to power loss reduction, frequency regulation to voltage support, customer privacy protection, and overall operational cost minimization [5].
- **Measurement and Communication Infrastructure**: Various functional measurement and communication infrastructures are necessary to ensure the proper operation of a community microgrid. These include IoT sensors, smart meters, data saving and logging devices, transceivers, programmable logic controllers (PLCs), internet communication networks, user-end interfaces, etc. These devices are gaining popularity across community microgrids due to their monitoring, control, and conditioning capacities. For instance, sensors and power meters for electrical parameters (current, voltage, frequency, energy consumption, power factor, and power generation) and meteorological parameters (solar irradiance, wind speed, temperature, and humidity) are processed digitally and transmitted to the human–machine interface (HMI). Meanwhile, control commands are input into HMI by central operators and individual customers and transmitted back to the microgrid system.

Community microgrids open up a new methodology for designing, constructing, and operating the power grid, utilizing DERs to obtain a more cost-effective, secure, and sustainable energy system. They offer residents and businesses resilience and environmental and economic benefits. They differ from the conventional microgrids in a few respects, summarized as follows [6]:

Feature	Conventional microgrids	Community microgrids
Size	Built for a single customer or a small number of adjacent households	Constructed for a relatively large number of customers within a whole substation grid area
DER installation	Usually installed behind the meter within the owner's property	Usually installed in front of the meter at small-to-medium scale
Resilience and capacity	Reduce energy imported from utilities for a single location or customer	Provides partial or full-scale electrical power to an entire community

2.2 Challenges and Key Technologies of Community Microgrids

Community microgrids come with many technological, safety, and regulatory challenges. Unlike conventional power networks, a community microgrid must provide sufficient energy to meet its local, perhaps moderated, demand. Besides, a community microgrid must be able to self-stabilize despite insufficient inertia within the system, particularly in islanded mode. This becomes more challenging with the increasing penetration level of power electronics-based electric appliances and DERs. Thus, intelligent power and energy management methods are necessary for the reliable and safe operation of DERs together with demand response. Further, the performance of a community microgrid depends on the effectiveness of cyber layers to a large extent. Unfortunately, measurement inaccuracy from sensors, data latency and dropout from communication networks, and cyberattacks from hackers create new vulnerabilities to the entire system. In addition to technological and safety challenges, regulatory challenges and security issues also need to be tackled as DERs within a community microgrid are operated by private customers. For instance, even with the availability of technologies, the interconnection rules with the main grid will probably take a while to achieve mutual agreement among various stakeholders, including individual customers, local councils, electricity distributors, etc. Concerns about tariffs, compensation, and environmental impact will probably become obstacles during the rollout of community microgrid projects.

The following highlights the key technologies to address these aforementioned challenges for community microgrids in relation to control, operation, and optimization:

- Intelligent Power Electronic Converters

Power converters, as the electronic interface between the DERs and the distribution power network, are the key electronic devices to achieve power conversion. They can be controlled to convert a voltage from one level to another with a specified amplitude and frequency within microgrids [7]. Regardless of their topologies, power electronic converters can be generally classified into three types, depending on their roles: grid-forming converters, grid-feeding converters, and grid-supporting converters. Maintaining stable voltage in islanded mode and delivering power to an energized grid in grid-connected mode are the primary tasks of grid-forming inverters and grid-feeding inverters, respectively, while grid-supporting inverters aim to provide support to the regulation of grid voltage and frequency. Unfortunately, industrially available converters mainly operate at grid-feeding modes. Besides, fault protection and fault ride-through capabilities of existing converters need to be re-designed to accommodate variations in fault current magnitudes during fault conditions in microgrids. The development of intelligent power converters to enable various operation modes will increase the flexibility and capability to work with stiff and weak grid conditions considering the expected high penetration of renewable energy sources.

Advanced control techniques, e.g., virtual synchronous generators, can be used to address power quality problems such as voltage harmonics, spikes, and frequency fluctuations. Various control techniques have been developed for singly operated power converters and multiple parallel converters, including cascaded linear control [8, 9], direct power control [10, 11], predictive control [12–14], virtual synchronous generators [15], concentrated control [16], instantaneous current-sharing method [17], droop strategy [18, 19], virtual oscillator control (VOC) [20, 21], etc. For high percentage integration of intermittent renewable energy sources and variable loads, advanced control algorithms with fast transient response and flexibility in considering different constraints are highly desired to enable grid-friendly plug-and-play integration of DERs.

- Dynamic Energy Management (demand response, generation scheduling, etc.)

To ensure the proper operation of a community microgrid, it is necessary to have a dynamic energy management system to achieve real-time generation and consumption coordination, thus regulating energy dispatch to fulfill various objectives, such as power loss reduction, operational cost minimization, etc. [22]. As mentioned in Sect. 2.2.1 an energy management system is essentially for optimal use of DERs in a coordinated, reliable, intelligent, and secure way. Such dynamic energy management includes demand response, generation scheduling, utility grid connection/ disconnection, etc. It considers both supply and demand-side management while satisfying system constraints to realize a sustainable, reliable, and economical operation [5, 23]. It offers benefits from energy savings to reactive power support, operational cost reduction to reduced emissions, and customer privacy protection to customer engagement. For example, when power demand tends to surpass generation, the dynamic energy management system may automatically and remotely shed non-critical loads within households and buildings to prevent voltage sags or even a blackout.

- Forecasting (generation forecasting and load forecasting)

Load and generation forecasting is vital in community microgrid operation [24, 25]. A forecasting algorithm is used to predict the demand for the loads in the microgrid network or the power generated by DERs in the near future. Forecasting helps to schedule the power optimally generated by DERs in advance. In the coming years, it is expected that artificial intelligence will likely be applied increasingly in generation and load forecasting in community microgrids.

- Fault Protection

The increasing penetration level of installed DERs in community microgrids causes variations in fault current magnitudes, which could deteriorate the reliability of conventional protection systems. In addition, increasingly complex and dynamically changing electrical networks due to the plug-and-play operations of DERs and the connection and disconnection of the community microgrid to the utility grid make the protection more challenging. It is noted that different sources can contribute to faults in a microgrid, including renewable generations, electric vehicles, or battery storage systems that are interfaced through power converters and transformers. Recently, it has been reported that the magnitudes of fault currents in a microgrid are usually much smaller than those in a conventional power grid, particularly in islanded modes [26, 27]. Therefore, conventional protection devices and the associated mechanisms should be re-designed to adapt to the new microgrid environment, and the protection must respond to different utility grid and microgrid faults.

- Information Communication (sensors, metering, data communication network, communication protocol, etc.)

With increasing smart meters installed at different locations of the power grid, large amounts of data are being generated, which will need to be collected, transferred, and processed. Thus, advanced information communication technologies, either wireless or cable-based, have become significant to cope with data and information explosion. A reliable and effective communication architecture includes physical hardware and communication protocols. Nowadays, to facilitate data and information communication among different key components within a community microgrid, the Internet of Things (IoT) has been deployed to integrate sensors, smart meters, DERs, loads, and customers into a digital highway [28].

- Sophisticated Interconnection Policies and Regulations

To manage DER penetration and integration, sophisticated regulations are needed as the power network in a community microgrid would cross property boundaries and contain many local customers. Also, interconnection standards between the microgrid and the main grid are required to facilitate the process and minimize the impacts on the main grid. Furthermore, stakeholders need generation, distribution, and electricity retail licenses to ensure a fair and competitive environment.

2.2.1 Opportunities and Recent Development

Community microgrids have found potential applications in different regions and countries with many local communities. For instance, while most of the population is concentrated in the capital cities in Australia, there are many other cities and towns across the broader areas. In this context, community microgrids are an effective solution for townships at the outer edges of the main electricity grid. Especially considering increasing extreme weather events and other natural disasters, community microgrids provide a more stable electricity supply and energy security. Over the past few years, supported by government, local councils, and industry, several community microgrid initiatives and projects have been put forward.

Back in 2010, a microgrid project was initiated on King Island in Tasmania where electricity supply is not available from the main power grid. Building on the 6 MW diesel fuel power station, this hybrid off-grid power system is now capable of providing 65% of energy needs on the island using renewable energy [29]. In 2016, Synergy initiated a project to trial a community-scale energy-serving model in north Perth. In collaboration with Alkimos Beach development partners Lendlease and Development WA, the project aimed to demonstrate the feasibility of providing electricity to households by connecting local individual solar PVs with a community-scale energy storage system [30]. A lesson learned in rural areas after the 2019–20 bushfires is to generate reliable electricity independently of the main utility grid during bushfires and extreme weather events. DELWP and AusNet are working with local councils and residents in Mallacoota, Omeo, and Corryong in Victoria to increase their bushfire and energy resilience through a Community Microgrids and Sustainable Energy Program [31]. In this program, solar power, diesel generators, batteries, and other energy infrastructure will be installed and interconnected to increase access to local generation and storage. From 2011 to 2023, as part of the *Regional and Remote Communities Reliability Fund Microgrids Program* funded by the Commonwealth Department of Industry, Science, Energy and Resources, C4NET coordinated the Donald and Tarnagulla Microgrid Feasibility Study [32]. The study was divided into 12 sub-projects conducted by various universities and industry partners across Victoria, including the Microgrid Islanding Design and Cost Analysis led by Federation University Australia. The project team analyzed 32 islanding scenarios for the local distribution network and performed a techno-economic analysis for each scenario. In NSW, a community microgrid project is being developed in two coastal towns of Bawley Point and Kioloa, with around 100 homes [33]. The community microgrid will be built as a self-sustaining energy system, harnessing electricity from rooftop solar and home-based batteries. Also, a 3 MW grid-connected battery is designed and positioned strategically between the two communities. This configuration enables the microgrid to operate autonomously and continuously supply electrical power during bushfires and storms.

References

1. N.M. Tabatabaei, *Microgrid Architectures, Control and Protection Methods* (Springer, 2020)
2. https://carbontrack.com.au/blog/community-microgrid/
3. J. de Hoog, T. Alpcan, M. Brazil, D.A. Thomas, I. Mareels, A market mechanism for electric vehicle charging under network constraints. IEEE Trans. Smart Grid **7**(2), 827–836 (2016). https://doi.org/10.1109/TSG.2015.2495181
4. J. de Hoog et al., Electric vehicle charging and grid constraints: comparing distributed and centralized approaches, in *2013 IEEE Power & Energy Society General Meeting, Vancouver, BC, Canada* (2013), pp. 1–5. https://doi.org/10.1109/PESMG.2013.6672222.
5. J. Hu et al., Economic model predictive control for microgrid optimization: a review. IEEE Trans. Smart Grid **15**(1), 472–484 (2024)
6. https://clean-coalition.org/community-microgrids/
7. J. Hu, Z. Li, J. Zhu, J.M. Guerrero, Voltage stabilization: a critical step toward high photovoltaic penetration. IEEE Ind. Electron. Mag. **13**(2), 17–30 (2019)
8. M. Abdel-Rahim, J.E. Quaicoe, Analysis and design of a multiple feedback loop control strategy for single-phase voltage-source UPS inverters. IEEE Trans. Power Electron. **11**(4), 532–541 (1996)
9. J. Dannehl, C. Wessels, F.W. Fuchs, Limitations of voltage-oriented PI current control of grid-connected PWM rectifciers with LCL filters. IEEE Trans. Ind. Electron. **56**(2), 380–388 (2009)
10. S. Kouro et al., Recent advances and industrial applications of multilevel converters. IEEE Trans. Ind. Electron. **57**(8), 2553–2580 (2010)
11. J. Hu, J. Zhu, D.G. Dorrell, In-depth study of direct power control strategies for power converters. IET Power Electron. **7**(7), 1810–1820 (2014)
12. J. Hu, J.M. Guerrero, S. Islam, *Model Predictive Control for Microgrids: From Power Electronic Converters to Energy Management* (IET, UK, 2021)
13. Y. Shan et al., A model predictive control for renewable energy based AC microgrids without any PID regulators. IEEE Trans. Power Electron. **33**(11), 9122–9126 (2018)
14. Y. Shan, J. Hu, H. Liu, A holistic power management strategy of microgrids based on Model Predictive Control and particle swarm optimization. IEEE Trans. Ind. Inform. (in press)
15. Q.C. Zhong, G. Weiss, Synchronverters: inverters that mimic synchronous generators. IEEE Trans. Ind. Electron. **58**(4), 1259–1267 (2011)
16. A.P. Martins, A.S. Carvalho, A.S. Araujo, Design and implementation of a current controller for the parallel operation of standard UPSs, in *Proceedings of the IEEE IECON* (1995), pp. 584–589
17. C.S. Lee et al., Parallel UPS with a instantaneous current sharing control. IEEE IECON (1998)
18. J.M. Guerrero, J.C. Vasquez, J. Matas, L.G. de Vicuna, M. Castilla, Hierarchical control of droop-controlled AC and DC microgrids—a general approach toward standardization. IEEE Trans. Ind. Electron. **58**(1), 158–172 (2011)

19. Zi. Li, K.W. Chan, J. Hu, Adaptive droop control using adaptive virtual impedance for microgrids with variable PV outputs and load demands. IEEE Trans. Ind. Electron. (2020)
20. B.B. Johnson, Synchronization of parallel single-phase inverters with virtual oscillator control. IEEE Trans. Power Electron. **29**(11), 6124–6138 (2014)
21. B.B. Johnson, Synchronization of nonlinear oscillators in an LTI electrical power network. IEEE Trans. Circ. Syst. (2014)
22. I. Mareels, J. de Hoog, D. Thomas, M. Brazil, T. Alpcan, D. Jayasuriya, V. Müenzel, L. Xia, R.R. Kolluri, On making energy demand and network constraints compatible in the last mile of the power grid. Ann. Rev. Control **38**(2), 243–258 (2014). ISSN 1367–5788
23. M.F. Zia, E. Elbouchikhi, M. Benbouzid, Microgrids energy management systems: a critical review on methods, solutions, and prospects. IEEE Ind. Electron. Mag. **13**(2), 17–30 (2019)
24. K. Zor, O. Timur, A. Teke, A state-of-the-art review of artificial intelligence techniques for short-term electric load forecasting, in *6th International Youth Conference on Energy* (2017)
25. J. Shi et al., Forecasting power output of photovoltaic systems based on weather classification and support vector machines. IEEE Ind. Appl. **48**(3), 17–30 (2012)
26. W.M. Hamanah et al., AC microgrid protection schemes: a comprehensive review. IEEE Access **11**, 76842–76868 (2023)
27. A. Anvari-Moghaddam, *Microgrids—Advances in Operation, Control, and Protection* (Springer, 2021). ISBN: 978-3-030-59749-8
28. D. Minoli, K. Sohraby, B. Occhiogrosso, IoT considerations, requirements, and architectures for smart buildings—energy optimization and next-generation building management systems. IEEE Internet Things J. **4**(1), 269–283 (2017)
29. https://www.hydro.com.au/clean-energy/hybrid-energy-solutions/success-stories/king-island
30. https://arena.gov.au/knowledge-bank/alkimos-beach-energy-storage-trial-final-report/
31. https://engage.vic.gov.au/community-microgrids
32. https://c4net.com.au/projects/donald-and-tarnagulla-microgrid-feasibility-study/
33. https://www.endeavourenergy.com.au/news/media-releases/first-community-microgrid-of-its-kind-in-nsw-on-track-to-deliver-energy-independence-for-residents

Chapter 3
Intelligent Power Electronic Converters and Control for Microgrids: A Master-Slave Model Predictive Approach

Fernanda Carnielutti, Mokhtar Aly, Margarita Norambuena, Jiefeng Hu, Josep Guerrero, and Jose Rodriguez

Abstract This book chapter presents Model Predictive Control (MPC) strategies for Master-Slave parallel inverters in microgrids. The Master is a grid-forming inverter with an LC output filter, while the Slave is a grid-following inverter with an output LCL filter. The microgrid is also composed of linear and non-linear loads, as well as grid and line impedances. First, a Finite Control Set MPC (FCS-MPC) is described and validated by means of Hardware-in-the-Loop (HIL) results covering different operational conditions of the microgrid, including grid connection, islanded mode, and load variations. Even though FCS-MPC has a very good dynamic and transient performance, it inherently results in variable switching frequency for the inverters, increasing the harmonic content of the output voltages and currents and making the design of the output filter more complex. In order to overcome these issues, a Fixed Switching Frequency Model Predictive Control (FSF-MPC) is then presented for the same microgrid, being again validated through HIL for the same operational conditions as for the FCS-MPC. Finally, the performance of both MPC controllers is compared. The results presented in this chapter demonstrate the good performances

F. Carnielutti (✉)
Federal University of Santa Maria, Santa Maria, Brazil
e-mail: fernanda.carnielutti@gmail.com

M. Aly · J. Rodriguez
Universidad San Sebastian, Santiago, Chile
e-mail: mokhtar.aly@uss.cl

J. Rodriguez
e-mail: jose.rodriguezp@uss.cl

M. Norambuena
Universidad Tecnica Federico Santa Maria, Valparaiso, Chile
e-mail: margarita.norambuena@usm.cl

J. Hu
Federation University, Ballarat, Australia
e-mail: j.hu@federation.edu.au

J. Guerrero
Aalborg University, Aalborg East, Denmark
e-mail: joz@energy.aau.dk

J. Hu (ed.), *Community Energy and Microgrids*, Green Energy and Technology,
https://doi.org/10.1007/978-981-97-6297-2_3

of the proposed FCS-MPC and FSF-MPC approaches, such as fast dynamic response, multi-variable control, robustness to parametric uncertainties, and fixed switching frequency for the FSF-MPC.

3.1 Introduction

Microgrids constitute systems generally composed of distributed generation units integrated with their own power electronics conversion systems, loads, distribution lines, management and control systems, relays, and communication technologies [1]. The control of a microgrid is usually implemented as a three-level hierarchical control [1–3]. The primary control is designed to operate in a time scale of milliseconds to seconds, and it is responsible for current and voltage control, virtual impedance emulation, islanding detection, etc. [4]. The secondary control acts in a time scale of seconds to minutes, performing voltage and frequency regulation, active and reactive power control, grid synchronization, etc. [5]. Finally, the tertiary control operates at a larger time, from minutes to hours, controlling the interaction of the microgrid to the main grid, including power flow regulation, market-driven optimal dispatch of the energy sources, etc.

Hierarchical control is commonly used for controlling microgrids, employing multi-loop cascaded linear control systems with modulation techniques including centralized or distributed control, average load distribution, master-slave based configurations, and droop mechanisms [1, 4, 6–8]. Specifically, concerning the utilization of Master-Slave based control within microgrids, various methodologies have been proposed in literature [9–11]. One approach has been presented using the robust controller with nested linear quadratic regulators (LQR) and mixed optimal control methods [12]. Additionally, a linear controller implementing cooperative load sharing has been introduced for managing multiple Master-Slave microgrids [13]. Another linear controller without a droop loop was presented in [14] for master-slave DG-based microgrids, which also eliminates explicit communications. Neural Networks (NNs) have also found applications in Master-Slave control of microgrids, such as the self-constructing based fuzzy NNs (SFNN) [15], data-driven three-layer NN [16]. Despite the prevalence of these strategies in both academic literature and practical implementations, they do exhibit certain drawbacks, such as inherent sluggish dynamic responses and challenges in implementing multi-faceted control objectives [1]. Even though these implementations are well-known and their use is widespread in real microgrid systems, they have some disadvantages, such as inherently slow dynamic responses and difficulties to implement multi-objective control [1]. These approaches also suffer from complex tuning procedures of the existing cascaded controllers, and/or large amounts of data for the tuning process, and/or unstable responses to transient events in microgrids under consideration.

An important alternative to traditional linear controllers and modulators is the utilization of Model Predictive Control (MPC), which is progressively gaining attention for the control of multiple parallel inverters within microgrids [1, 5, 6]. Among the most common MPC methodologies, Finite Control Set MPC (FCS-MPC) represents a suitable candidate for controlling microgrid systems. In FCS-MPC methods, the discrete-time model of the system is employed for predicting its forthcoming states through the integration of system inputs and dynamic equations [17–19]. During each sampling instance, a cost function is evaluated for all system inputs. Then, the selection of the switching state that minimizes the cost function is made and its corresponding switching pulses are applied to the power inverter system. The design of the cost function can encompass diverse control objectives, accommodating system restrictions, limitations, and non-linearities as well. As a result, MPC has the capacity to incorporate the operational boundaries of the system within its formulation, leading to faster and smoother dynamic responses when compared to conventional linear controllers [6].

Consequently, the conventional FCS-MPC technique has demonstrated its efficacy for grid-tied inverters and microgrid scenarios. In this context, MPC approaches can be implemented at the converter level and/or grid level, both for single-phase and three-phase systems. The converter-level MPC is responsible for the primary control of the inverters in the microgrid, that is, current and power control, capacitor voltage control for multilevel inverters, emulation of virtual impedance for parallel inverters, operation on islanded mode, etc. Numerous studies have addressed primary control, such as the works presented in [20–29].

On the other hand, as the name implies, the grid-level MPC is responsible for performing secondary and tertiary controls [2, 5]. For example, a centralized MPC was proposed in [30], capable of performing load frequency control in multi-area interconnected power systems. A distributed MPC was presented in [31] for automatic generation control of interconnected microgrids. Another distributed MPC was proposed in [32], that performs voltage and frequency control and power sharing in autonomous microgrids. Another MPC method was proposed and described in [33] to restore the frequency in microgrids, considering communication delays.

With respect to tertiary control, in [34], a two-layer MPC was proposed for performing an optimal operation of a microgrid comprised of photovoltaic (PV) systems, batteries, and a diesel genset. The first optimization layer performs optimal energy dispatch, and the second one adjusts the power of the diesel genset. An MPC for optimized operation of a microgrid was presented in [35] that is able to satisfy time-varying requests and operational constraints. In [36], a sampling-based MPC that uses real-time price of energy and forecasts of PV and load power was proposed so as to optimize the power dispatch of a PV-Battery ESS (PV-BESS) microgrid and minimize overall costs. Also for PV-BESS systems, in [37], an MPC is presented for the inner loop of a bidirectional buck-boost converter for the BESS unit of the microgrid. In this work, the parallel operation of power converters is performed by another MPC with a droop control, synthesizing the voltages for the microgrid and performing load sharing.

Also, MPC algorithms have found extensive utilization in several microgrid applications [38–40]. In [41], an MPC-based method was proposed for load sharing in microgrids and power dispatch of distributed generation systems. The presented method is based on MPC incorporated with a droop controller, namely the holistic power controller. Another model predictive power controller (MPPC) was integrated with the economic dispatch for optimizing power flow in microgrid systems in [42]. A fast restoration method for voltage and frequency in the interconnected microgrids was proposed in [43]. Another robust fuzzy MPC method was proposed in [44] for controlling secondary voltage in islanded microgrid systems. In these published papers, the MPC has shown a flexible solution for controlling current and/or voltage by including multiple objectives in a single cost function. Thence, MPC is capable of avoiding the cascaded control structures, which require complex design procedures.

Nevertheless, the classical FCS-MPC methodology has certain inherent challenges, such as widespread harmonic spectra in the output voltages and currents and variable switching frequencies. To address these concerns, the literature has proposed MPC strategies that result in fixed switching frequencies in the output variables [21, 23, 27, 45, 46], and that will be addressed here as Fixed Switching Frequency MPC (FSF-MPC). Even though many interesting and effective FSF-MPC strategies have been presented in the literature, the widespread integration of these MPC algorithms within microgrid applications has been very limited.

Within this context, this chapter presents first a Finite Control Set MPC (FCS-MPC) for Master-Slave parallel inverters in microgrids. The Master component constitutes a grid-forming two-level inverter, incorporating an output LC filter under the governance of a Voltage MPC. On the other hand, the Slave component takes the form of a grid-following two-level inverter, integrated with an output LCL filter and directed by a Current MPC. The microgrid configuration encompasses both linear and non-linear loads, with the model incorporating the grid and line impedances. The proposed FCS-MPC is described and validated by means of Hardware-in-the-Loop (HIL) results covering different operational conditions of the microgrid, including grid connection, islanded mode, and load variations. This FCS-MPC is based on the work presented in [47]. Even though FCS-MPC has a very good dynamic and transient performance, as previously discussed, it inherently results in variable switching frequency for the inverters, increasing the harmonic content of the output voltages and currents and making the design of the output filter more complex.

In order to overcome these issues, this chapter also proposes a Fixed Switching Frequency Model Predictive Control (FSF-MPC), presented for the same microgrid, being again validated through HIL for the same operational conditions as for the FCS-MPC. The FSF-MPC framework advocated herein is an expansion upon the preceding FCS-MPC-based Master-Slave method, which exhibited variable switching frequencies in [47]. Additionally, the approach ensures a fixed switching frequency, bringing together the merits of classical MPC and linear control approaches. Finally, the performance of both MPC controllers is compared. The results presented in this chapter demonstrate the good performances of the proposed FCS-MPC and FSF-MPC approaches, such as fast dynamic response, multi-variable control, robustness to parametric uncertainties, and fixed switching frequency for the FSF-MPC, evaluating and proving its performance across diverse operational conditions within the microgrid environment.

3.2 Proposed Finite Control Set MPC (FCS-MPC) for Master-Slave Inverters

The microgrid considered in this application is shown in Fig. 3.1, and the block diagram of the FCS-MPC is depicted in Fig. 3.2. The Master is a grid-forming two-level inverter with an output LC filter and Voltage MPC, while the Slave is a grid-following two-level inverter with an output LCL filter and Current MPC. In this work, the DC-inputs of both inverters are modeled as ideal DC sources; however, in a real application, the DC-input of the Master inverter could be a Battery Energy Storage System (BESS), and of the Slave inverter a photovoltaic panel (PV), for example.

In the following sections, the FCS-MPC for both the Master and Slave inverters will be presented.

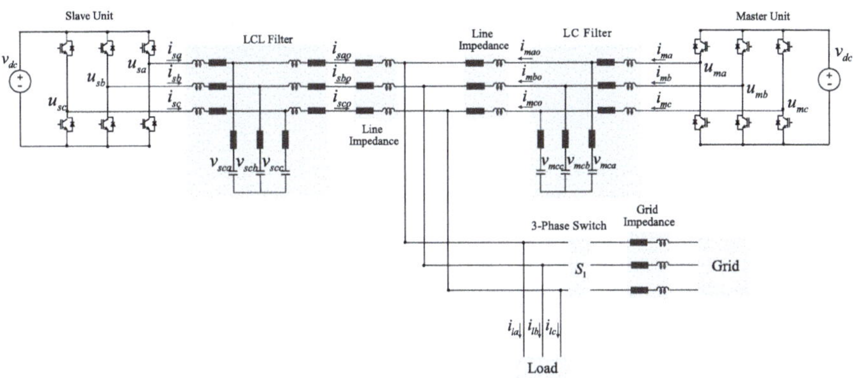

Fig. 3.1 Schematic of the microgrid considered in this application

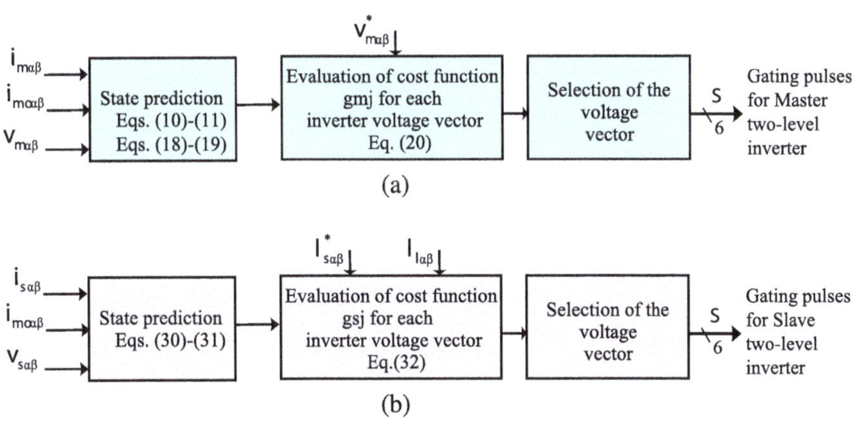

Fig. 3.2 Block diagram of the proposed Master-Slave FCS-MPC: **a** Proposed FCS-MPC for Master inverter; **b** Proposed FCS-MPC for Slave inverter

3.2.1 Voltage FCS-MPC for the Master Inverter

First, the Voltage FCS-MPC for the Master inverter will be presented. As previously mentioned, the Master is a grid-forming two-level inverter with an output LC filter, whose main objective is to synthesize voltages with adequate amplitudes and frequencies for the microgrid. In order to describe the proposed Voltage FCS-MPC, let us begin by writing the continuous-time dynamic equations of the inverter-side currents of the Master inverter in *abc* coordinates

$$L_m\left(\frac{di_{ma}(t)}{dt}-\frac{di_{mb}(t)}{dt}\right)=R\left(i_{ma}(t)-i_{mb}(t)\right)+u_{ma}(t)-u_{mb}(t)-v_{ma}(t)+v_{mb}(t) \tag{3.1}$$

$$L_m\left(\frac{di_{mb}(t)}{dt}-\frac{di_{mc}(t)}{dt}\right)=R\left(i_{mb}(t)-i_{mc}(t)\right)+u_{mb}(t)-u_{mc}(t)-v_{mb}(t)+v_{mc}(t) \tag{3.2}$$

where i_{ma}, i_{mb} and i_{mc} are the currents of the inverter-side inductor of the LC filter, v_{mca}, v_{mcb} and v_{mcc} are the voltages of the filter capacitors and L_m and R are, respectively, the filter inductance and its associated resistance. As the microgrid of Fig. 3.1 is a three-phase three-wire system, we have that $i_{ma}+i_{mb}+i_{mc}=0$, and thus i_{mc} is linearly dependent on i_{ma} and i_{mb}.

In state space, we can write (3.1) and (3.2) as

$$L_m\begin{bmatrix}1&-1&0\\0&1&-1\end{bmatrix}\begin{bmatrix}\frac{di_{ma}(t)}{dt}\\\frac{di_{mb}(t)}{dt}\\\frac{di_{mc}(t)}{dt}\end{bmatrix}=R\begin{bmatrix}1&-1&0\\0&1&-1\end{bmatrix}\begin{bmatrix}i_{ma}(t)\\i_{mb}(t)\\i_{mc}(t)\end{bmatrix}+\begin{bmatrix}1&-1&0\\0&1&-1\end{bmatrix}\begin{bmatrix}u_{ma}(t)\\u_{mb}(t)\\u_{mc}(t)\end{bmatrix}-\begin{bmatrix}1&-1&0\\0&1&-1\end{bmatrix}\begin{bmatrix}v_{ma}(t)\\v_{mb}(t)\\v_{mc}(t)\end{bmatrix} \tag{3.3}$$

The model presented in (3.3) is a coupled system. In order to decouple it, the Clark transformation can be applied and the variables can be written in $\alpha\beta$ coordinates as

$$\begin{bmatrix}i_{m\alpha}(t)\\i_{m\beta}(t)\end{bmatrix}=\frac{2}{3}\begin{bmatrix}1&-\frac{1}{2}&\frac{1}{2}\\0&\frac{\sqrt{3}}{2}&\frac{\sqrt{3}}{2}\end{bmatrix}\begin{bmatrix}i_{ma}(t)\\i_{mb}(t)\\i_{mc}(t)\end{bmatrix} \tag{3.4}$$

$$\begin{bmatrix}u_{m\alpha}(t)\\u_{m\beta}(t)\end{bmatrix}=\frac{2}{3}\begin{bmatrix}1&-\frac{1}{2}&\frac{1}{2}\\0&\frac{\sqrt{3}}{2}&\frac{\sqrt{3}}{2}\end{bmatrix}\begin{bmatrix}u_{ma}(t)\\u_{mb}(t)\\u_{mc}(t)\end{bmatrix} \tag{3.5}$$

$$\begin{bmatrix} v_{m\alpha}(t) \\ v_{m\beta}(t) \end{bmatrix} = \frac{2}{3} \begin{bmatrix} 1 & -\frac{1}{2} & \frac{1}{2} \\ 0 & \frac{\sqrt{3}}{2} & \frac{\sqrt{3}}{2} \end{bmatrix} \begin{bmatrix} v_{ma}(t) \\ v_{mb}(t) \\ v_{mc}(t) \end{bmatrix} \tag{3.6}$$

where $i_{m\alpha\beta}$ are the inverter-side currents on the inductor of the LC filter, $u_{m\alpha\beta}$ are the PWM phase voltages of the Master inverter and $v_{m\alpha\beta}$ are the voltages of the filter capacitors. As a result, the continuous-time dynamic equations of the inverter-side currents of the Master inverter in $\alpha\beta$ coordinates are

$$L_m \begin{bmatrix} 1 & 0 \\ 0 & 1 \end{bmatrix} \begin{bmatrix} \frac{di_{m\alpha}(t)}{dt} \\ \frac{di_{m\beta}(t)}{dt} \end{bmatrix} = R \begin{bmatrix} 1 & 0 \\ 0 & 1 \end{bmatrix} \begin{bmatrix} i_{m\alpha}(t) \\ i_{m\beta}(t) \end{bmatrix} + \begin{bmatrix} 1 & 0 \\ 0 & 1 \end{bmatrix} \begin{bmatrix} u_{m\alpha}(t) \\ u_{m\beta}(t) \end{bmatrix} - \begin{bmatrix} 1 & 0 \\ 0 & 1 \end{bmatrix} \begin{bmatrix} v_{m\alpha}(t) \\ v_{m\beta}(t) \end{bmatrix} \tag{3.7}$$

As the FCS-MPC will be implemented in a Digital Signal Processor (DSP) or microprocessor, the continuous-time model presented in 3.7 needs to be discretized. In order to do so, the Euler discretization method can be applied, where the derivatives of the currents can be expressed as

$$\frac{di_{m\alpha}(t)}{dt} = \frac{i_{m\alpha}(k+1) - i_{m\alpha}(k)}{T_s} \tag{3.8}$$

$$\frac{di_{m\beta}(t)}{dt} = \frac{i_{m\beta}(k+1) - i_{m\beta}(k)}{T_s} \tag{3.9}$$

where T_s is the sampling period.

Then, the discrete-time dynamic equations of the inverter-side currents of the Master inverter in $\alpha\beta$ coordinates are

$$i_{m\alpha}(k+1) = \left(1 - \frac{RT_s}{L_m}\right) i_{m\alpha}(k) + \frac{T_s}{L_m} u_{m\alpha}(k) - \frac{T_s}{L_m} v_{m\alpha}(k) \tag{3.10}$$

$$i_{m\beta}(k+1) = \left(1 - \frac{RT_s}{L_m}\right) i_{m\beta}(k) + \frac{T_s}{L_m} u_{m\beta}(k) - \frac{T_s}{L_m} v_{m\beta}(k) \tag{3.11}$$

As the controlled variables of the FCS-MPC for the Master inverter are the voltages of the capacitors of the LC filter, let us now derive the dynamic equations of these variables. The continuous-time dynamic equations of the voltages of the capacitors of the LC filter of the Master inverter in *abc* coordinates are

$$v_{ma}(t) = v_{ma}(0) + \frac{1}{C_m} \int_0^t (i_{ma}(\tau) - i_{mao}(\tau)) d\tau \tag{3.12}$$

$$v_{mb}(t) = v_{mb}(0) + \frac{1}{C_m} \int_0^t (i_{mb}(\tau) - i_{mbo}(\tau)) d\tau \tag{3.13}$$

$$v_{mc}(t) = v_{mc}(0) + \frac{1}{C_m}\int_0^t (i_{mc}(\tau) - i_{mco}(\tau))d\tau \tag{3.14}$$

where C_m is the filter capacitance and i_{mao}, i_{mbo} and i_{mco} are the output currents of the Master inverter, measured after the LC filter and considered here as disturbances. In order to represent the capacitor voltages in the $\alpha\beta$ coordinates, the currents i_{mao}, i_{mbo} and i_{mco} are transformed as

$$\begin{bmatrix} i_{mo\alpha}(t) \\ i_{mo\beta}(t) \end{bmatrix} = \frac{2}{3}\begin{bmatrix} 1 & -\frac{1}{2} & \frac{1}{2} \\ 0 & \frac{\sqrt{3}}{2} & \frac{\sqrt{3}}{2} \end{bmatrix}\begin{bmatrix} i_{mao}(t) \\ i_{mbo}(t) \\ i_{mco}(t) \end{bmatrix} \tag{3.15}$$

As a result, the voltages of the filter capacitors of the Master inverter in $\alpha\beta$ coordinates are

$$v_{m\alpha}(t) = v_{m\alpha}(0) + \frac{1}{C_m}\int_0^t (i_{m\alpha}(\tau) - i_{mo\alpha}(\tau))d\tau \tag{3.16}$$

$$v_{m\beta}(t) = v_{m\beta}(0) + \frac{1}{C_m}\int_0^t (i_{m\beta}(\tau) - i_{mo\beta}(\tau))d\tau \tag{3.17}$$

Again, the Euler discretization method is used to discretize the system, and the discrete-time dynamic equations of the filter capacitor voltages of the Master inverter in $\alpha\beta$ coordinates as

$$v_{m\alpha}(k+1) = \frac{T_s}{C_m}(i_{m\alpha}(k) - i_{mo\alpha}(k)) + v_\alpha(k) \tag{3.18}$$

$$v_{m\beta}(k+1) = \frac{T_s}{C_m}\left(i_{m\beta}(k) - i_{mo\beta}(k)\right) + v_\beta(k) \tag{3.19}$$

where $i_{mo\alpha}$ and $i_{mo\beta}$ are the output currents of the Master inverter.

With the discrete-time equations of the system, it is possible now to move forward to the implementation of the proposed Voltage FCS-MPC for the Master inverter. In order to implement the Voltage FCS-MPC algorithm, first, for each sampling period T_s, the voltages $v_{m\alpha\beta}(k+1)$ are predicted for all inverter voltage vectors, i.e., eight for a three-phase two-level inverter, considering the two redundancies of the null vector. As the objective of the Master inverter is to control the voltages of the LC filter capacitors, the cost function for its Voltage FCS-MPC can be written as the error between the voltage reference $v^*_{m\alpha\beta}(k+1)$ and the predicted voltages as

$$g_{mj} = \left[(v^*_{m\alpha}(k+1) - v_{m\alpha j}(k+1))^2 + (v^*_{m\beta}(k+1) - v_{m\beta j}(k+1))^2\right] \tag{3.20}$$

where $j = 1 \ldots 8$ represents each inverter voltage vector. Then, during each T_s, the cost function in (3.20) is evaluated for all eight inverter voltage vectors, and the one that results in the lowest value of (3.20) is chosen and implemented by the Master inverter.

3.2.2 *Current FCS-MPC for the Slave Inverter*

For the Slave inverter, the procedure for its Current FCS-MPC is very similar to the Voltage FCS-MPC presented for the Master inverter in Sect. 3.2.1. As mentioned, the Slave is a grid-following inverter with LCL filter, whose objective is to control the output currents injected into the grid or the load. In this application, the resonant frequency of the LCL filter is attenuated by means of passive damping, i.e., the damping resistors in series with the filter capacitors in Fig. 3.1. To describe the proposed Current FCS-MPC, let us begin by writing the continuous-time dynamic equations of the inverter-side currents of the Slave inverter in *abc* coordinates

$$L_s \left(\frac{di_{sa}(t)}{dt} - \frac{di_{sb}(t)}{dt} \right) = R\left(i_{sa}(t) - i_{sb}(t)\right) + u_{sa}(t) - u_{sb}(t) - v_{sa}(t) + v_{sb}(t) \tag{3.21}$$

$$L_s \left(\frac{di_{sb}(t)}{dt} - \frac{di_{sc}(t)}{dt} \right) = R\left(i_{sb}(t) - i_{sc}(t)\right) + u_{sb}(t) - u_{sc}(t) - v_{sb}(t) + v_{sc}(t) \tag{3.22}$$

where i_{sa}, i_{sb} and i_{sc} are currents of the inverter-side inductor of the LCL filter, v_{sca}, v_{scb} and v_{scc} are the voltages of the filter capacitors and L_s and R are, respectively, the filter inductance and its associated resistance. As for the Master, the Slave inverter is a three-phase three-wire system, where $i_{sa} + i_{sb} + i_{sc} = 0$, and thus i_{sc} is linearly dependent on i_{sa} and i_{sb}.

Expressing (3.21) and (3.22) in state space, we have

$$L_s \begin{bmatrix} 1 & -1 & 0 \\ 0 & 1 & -1 \end{bmatrix} \begin{bmatrix} \frac{di_{sa}(t)}{dt} \\ \frac{di_{sb}(t)}{dt} \\ \frac{di_{sc}(t)}{dt} \end{bmatrix} = R \begin{bmatrix} 1 & -1 & 0 \\ 0 & 1 & -1 \end{bmatrix} \begin{bmatrix} i_{sa}(t) \\ i_{sb}(t) \\ i_{sc}(t) \end{bmatrix} + \begin{bmatrix} 1 & -1 & 0 \\ 0 & 1 & -1 \end{bmatrix} \begin{bmatrix} u_{sa}(t) \\ u_{sb}(t) \\ u_{sc}(t) \end{bmatrix} - \begin{bmatrix} 1 & -1 & 0 \\ 0 & 1 & -1 \end{bmatrix} \begin{bmatrix} v_{sa}(t) \\ v_{sb}(t) \\ v_{sc}(t) \end{bmatrix} \tag{3.23}$$

The model shown in (3.23) is a coupled. To have a decoupled system, the Clark transformation will be applied and the Slave variables in $\alpha\beta$ coordinate as

$$\begin{bmatrix} i_{s\alpha}(t) \\ i_{s\beta}(t) \end{bmatrix} = \frac{2}{3}\begin{bmatrix} 1 & -\frac{1}{2} & \frac{1}{2} \\ 0 & \frac{\sqrt{3}}{2} & \frac{\sqrt{3}}{2} \end{bmatrix}\begin{bmatrix} i_{sa}(t) \\ i_{sb}(t) \\ i_{sc}(t) \end{bmatrix} \tag{3.24}$$

$$\begin{bmatrix} u_{s\alpha}(t) \\ u_{s\beta}(t) \end{bmatrix} = \frac{2}{3}\begin{bmatrix} 1 & -\frac{1}{2} & \frac{1}{2} \\ 0 & \frac{\sqrt{3}}{2} & \frac{\sqrt{3}}{2} \end{bmatrix}\begin{bmatrix} u_{sa}(t) \\ u_{sb}(t) \\ u_{sc}(t) \end{bmatrix} \tag{3.25}$$

$$\begin{bmatrix} v_{s\alpha}(t) \\ v_{s\beta}(t) \end{bmatrix} = \frac{2}{3}\begin{bmatrix} 1 & -\frac{1}{2} & \frac{1}{2} \\ 0 & \frac{\sqrt{3}}{2} & \frac{\sqrt{3}}{2} \end{bmatrix}\begin{bmatrix} v_{sa}(t) \\ v_{sb}(t) \\ v_{sc}(t) \end{bmatrix} \tag{3.26}$$

where $i_{s\alpha\beta}$ are the inverter-side currents of the LCL filter, $u_{s\alpha\beta}$ are the PWM phase voltages of the Slave inverter and $v_{s\alpha\beta}$ are the voltages of the filter capacitors. As a result, the continuous-time dynamic equations of the inverter-side currents of the Slave inverter in $\alpha\beta$ coordinates are

$$L_s\begin{bmatrix} 1 & 0 \\ 0 & 1 \end{bmatrix}\begin{bmatrix} \frac{di_{s\alpha}(t)}{dt} \\ \frac{di_{s\beta}(t)}{dt} \end{bmatrix} = R\begin{bmatrix} 1 & 0 \\ 0 & 1 \end{bmatrix}\begin{bmatrix} i_{s\alpha}(t) \\ i_{s\beta}(t) \end{bmatrix} + \begin{bmatrix} 1 & 0 \\ 0 & 1 \end{bmatrix}\begin{bmatrix} u_{s\alpha}(t) \\ u_{s\beta}(t) \end{bmatrix} - \begin{bmatrix} 1 & 0 \\ 0 & 1 \end{bmatrix}\begin{bmatrix} v_{s\alpha}(t) \\ v_{s\beta}(t) \end{bmatrix} \tag{3.27}$$

The FCS-MPC for the Slave inverter will also be implemented in a Digital Signal Processor (DSP) or microprocessor. Consequently, the continuous-time model presented in (3.27) will be discretized by means of the Euler discretization method, where the derivatives of the currents are

$$\frac{di_{s\alpha}(t)}{dt} = \frac{i_{s\alpha}(k+1) - i_{s\alpha}(k)}{T_s} \tag{3.28}$$

$$\frac{di_{s\beta}(t)}{dt} = \frac{i_{s\beta}(k+1) - i_{s\beta}(k)}{T_s} \tag{3.29}$$

where T_s is the sampling period.

Then, the discrete-time dynamic equations of the inverter-side currents of the Slave inverter in $\alpha\beta$ coordinates are

$$i_{s\alpha}(k+1) = \left(1 - \frac{RT_s}{L_s}\right) i_{s\alpha}(k) + \frac{T_s}{L_s} u_{s\alpha}(k) - \frac{T_s}{L_s} v_{s\alpha}(k) \tag{3.30}$$

$$i_{s\beta}(k+1) = \left(1 - \frac{RT_s}{L_s}\right) i_{s\beta}(k) + \frac{T_s}{L_s} u_{s\beta}(k) - \frac{T_s}{L_s} v_{s\beta}(k) \tag{3.31}$$

where $v_{s\alpha}$ and $v_{s\beta}$ are the filter capacitors voltages of and $i_{s\alpha}$ and $i_{s\beta}$ are the inverter-side currents of the Slave inverter.

With the discrete-time equations of the system, let us now implement the proposed Current FCS-MPC for the Slave inverter. For each sampling period T_s, the inverter-side currents $i_{s\alpha\beta}(k+1)$ are predicted for all inverter voltage vectors. As the Slave is also a three-phase two-level inverter, it has eight voltage vectors, including the two redundancies of the null vector. As the objective of the Slave inverter is to control the currents (and, consequently, the active and reactive powers) injected into the grid, its cost function can be written as the error between the inverter-side current references $i^*_{s\alpha\beta}(k+1)$ and the predicted ones as

$$g_{sj} = \left[(i^*_{s\alpha}(k+1) - i_{s\alpha j}(k+1))^2 + (i^*_{s\beta}(k+1) - i_{s\beta j}(k+1))^2\right] \tag{3.32}$$

where $j = 1 \ldots 8$ represents each inverter voltage vector.

If the Slave is connected to the main grid, its power references, and, consequently, the current references $i^*_{s\alpha\beta g}(k)$ for the grid-side inductor of the LCL filter, can be defined from an outer controller; if the microgrid is islanded, these current references are calculated directly from the load currents. However, as shown in (3.32), the currents to be controlled by the Current FCS-MPC are the inverter-side ones, whose references are $i^*_{s\alpha\beta}(k)$. These can be calculated from the grid-side current references $i^*_{s\alpha\beta g}(k)$, compensating the reactive power of the filter as

$$i^*_{s\alpha}(k) = i^*_{s\alpha g}(k) - 2\pi f C_s v_{s\beta}(k) \tag{3.33}$$

$$i^*_{s\beta}(k) = i^*_{s\beta g}(k) + 2\pi f C_s v_{s\alpha}(k) \tag{3.34}$$

where f is the grid frequency and C_s is the LCL filter capacitance. The current references are then extrapolated to $(k+1)$ as

$$i^*_{s\alpha}(k+1) = i^*_{s\alpha}(k)\cos(\omega T_s) - i^*_{s\beta}(k)\sin(\omega T_s) \tag{3.35}$$

$$i^*_{s\beta}(k+1) = i^*_{s\beta}(k)\cos(\omega T_s) + i^*_{s\alpha}(k)\sin(\omega T_s) \tag{3.36}$$

where the grid frequency $\omega = 2\pi f$ comes from a Synchronous Reference Frame Phase Locked Loop (SRF-PLL) [48]. It must be stressed here that the grid-side currents of the LCL filter can also be controlled, as well as the voltages of the capacitors in order to accomplish active damping of the LCL filter resonance. However, the control of these variables is more difficult to implement when compared to the control of the inverter-side currents with passive damping of the resonant frequency of the LCL filter.

Finally, during each T_s, the cost function in (3.32) is evaluated for all eight inverter voltage vectors, and the one that results in the lowest value of (3.32) is chosen and implemented by the Slave inverter.

Table 3.1 System parameters for the case studies

Parameter	Value
Total DC-link voltage (Master and Slave)	800 V
Rated power (Master and slave)	100 kVA
LCL filter (Slave)	3 mH; 300 μF; 200 μH
LC filter (Master)	3 mH; 800 μF
Filter parasitic resistance (Master and Slave)	0.012 Ω
Filter damping resistor Master	1 mΩ
Filter damping resistor Slave	0.2 Ω
Grid impedance	$Z_g = 2.7\,\mu$H, 0.01 Ω
Line impedance	$Z_l = 20\,\mu$H, 0.01 Ω
Grid voltage	220 Vrms; 50 Hz
Sampling frequency f_s	20 kHz

3.3 Case Studies for the Master-Slave FCS-MPC

To validate the proposed Master-Slave FCS-MPC, the microgrid of Fig. 3.1 was implemented in a Typhoon HIL 604 Hardware-in-the-Loop (HIL) platform. The system parameters are shown in Table 3.1. The Voltage and Current MPCs, as well as other algorithms such as the SRF-PLL, were implemented in the C Block of Typhoon HIL, that includes the processor implementation delay. As a result, the cost functions are calculated for the predicted variables at $k+2$. First, the inverter-side currents of the Master inverter in $\alpha\beta$ coordinates at $k+2$ are

$$i_{m\alpha}(k+2) = \left(1 - \frac{RT_s}{L_s}\right) i_{m\alpha}(k+1) + \frac{T_s}{L_m} u_{m\alpha}(k+1) - \frac{T_s}{L_m} v_{m\alpha}(k+1) \quad (3.37)$$

$$i_{m\beta}(k+2) = \left(1 - \frac{RT_s}{L_m}\right) i_{m\beta}(k+1) + \frac{T_s}{L_m} u_{m\beta}(k+1) - \frac{T_s}{L_m} v_{m\beta}(k+1) \quad (3.38)$$

Also, the voltages of the capacitors of the LC filter in $k+2$ are

$$v_{m\alpha}(k+2) = \frac{T_s}{C_m}(i_{m\alpha}(k+1) - i_{mo\alpha}(k+1)) + v_\alpha(k+1) \quad (3.39)$$

$$v_{m\beta}(k+2) = \frac{T_s}{C_m}(i_{m\beta}(k+1) - i_{mo\beta}(k+1)) + v_\beta(k+1) \quad (3.40)$$

As can be seen from (3.37)–(3.38), the inverter-side currents $i_{m\alpha\beta}$ in $k+2$ are dependent on the inverter voltages $u_{m\alpha\beta}$ in $k+1$, that compose the voltage vector that will be synthesized by the inverter in T_s in order to control the voltages of the capacitors of the LC filter. Consequently, in order for $u_{m\alpha\beta}(k+2)$ to appear in the expressions of the capacitor voltages by means of $i_{m\alpha\beta}(k+2)$, these need to be further predicted to $k+3$ as

$$v_{m\alpha}(k+3) = \frac{T_s}{C_m}(i_{m\alpha}(k+2) - i_{mo\alpha}(k+2)) + v_\alpha(k+2) \tag{3.41}$$

$$v_{m\beta}(k+3) = \frac{T_s}{C_m}(i_{m\beta}(k+2) - i_{mo\beta}(k+2)) + v_\beta(k+2) \tag{3.42}$$

and, consequently, the voltage references are also extrapolated to $k+3$ as

$$v^*_{m\alpha}(k+2) = v^*_{m\alpha}(k+1)\cos(\omega T_s) - v^*_{m\beta}(k+1)\sin(\omega T_s) \tag{3.43}$$

$$v^*_{m\beta}(k+2) = v^*_{m\beta}(k+1)\cos(\omega T_s) + v^*_{m\alpha}(k+1)\sin(\omega T_s) \tag{3.44}$$

$$v^*_{m\alpha}(k+3) = v^*_{m\alpha}(k+2)\cos(\omega T_s) - v^*_{m\beta}(k+2)\sin(\omega T_s) \tag{3.45}$$

$$v^*_{m\beta}(k+3) = v^*_{m\beta}(k+2)\cos(\omega T_s) + v^*_{m\alpha}(k+2)\sin(\omega T_s) \tag{3.46}$$

Finally, the inverter-side currents of the Slave inverter in $\alpha\beta$ coordinates at $k+2$ are

$$i_{s\alpha}(k+2) = \left(1 - \frac{RT_s}{L_s}\right) i_{s\alpha}(k+1) + \frac{T_s}{L_s}u_{s\alpha}(k+1) - \frac{T_s}{L_s}v_{s\alpha}(k+1) \tag{3.47}$$

$$i_{s\beta}(k+2) = \left(1 - \frac{RT_s}{L_s}\right) i_{s\beta}(k+1) + \frac{T_s}{L_s}u_{s\beta}(k+1) - \frac{T_s}{L_s}v_{s\beta}(k+1) \tag{3.48}$$

and the inverter-side current references at $k+2$ are

$$i^*_{s\alpha}(k+2) = i^*_{s\alpha}(k+1)\cos(\omega T_s) - i^*_{s\beta}(k+1)\sin(\omega T_s) \tag{3.49}$$

$$i^*_{s\beta}(k+2) = i^*_{s\beta}(k+1)\cos(\omega T_s) + i^*_{s\alpha}(k+1)\sin(\omega T_s) \tag{3.50}$$

Let us now present the HIL case studies. First, the microgrid is connected to the main grid with a balanced three-phase RL load with $R_l = 1.03\,\Omega$ and $L_l = 3.33\,\text{mH}$. Then, the switch that connects it to the main grid is opened, and the microgrid is islanded. In islanded mode, the RL load is first changed to a three-phase unbalanced RL load with $R_l = 1.03\,\Omega$ and $L_l = 3.33\,\text{mH}$ for phase a, $R_l = 4\,\Omega$ and $L_l = 10\,\text{mH}$ for phase b and $R_l = 2\,\Omega$ and $L_l = 6\,\text{mH}$ for phase c. Finally, the load is changed to a non-linear load composed of a three-phase diode rectifier with an input L filter with $L = 1\,\text{mH}$ and an output RL load with $R = 5\,\Omega$ and $L = 5\,\text{mH}$. These results are shown in Fig. 3.3, where we can see, from top to bottom in *abc* coordinates: (i) Capacitor voltages of the LC filter of the Master inverter; (ii) grid-side currents of the Master and Slave inverters; (iii) total active power reference (coming from the

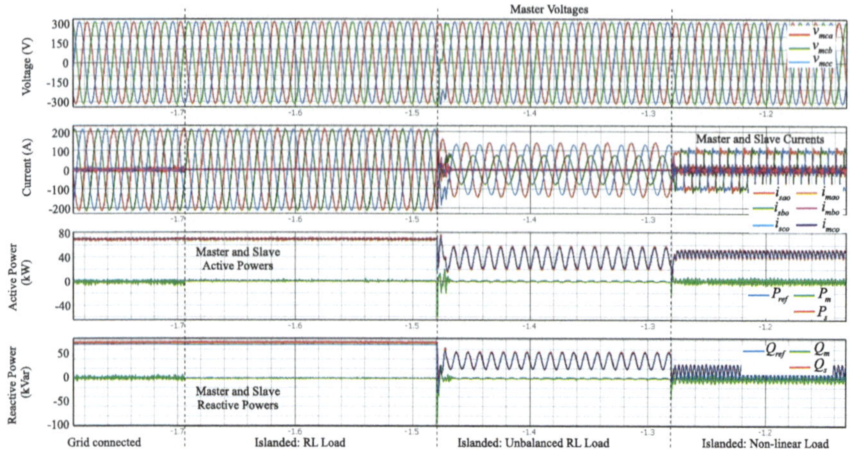

Fig. 3.3 HIL results for the FCS-MPC: (i) capacitor voltages of the LC filter of the Master inverter; (ii) grid-side currents of the Master and Slave inverters; (iii) total active power reference, and Master and Slave active powers; (iv) total reactive power reference, and Master and Slave reactive powers

load demand), and Master and Slave active powers; (iv) total reactive power reference (coming from the load demand), and Master and Slave reactive powers. Also, Figs. 3.4–3.6 show zooms in the transient results for, respectively, grid-connected to islanded operation with balanced RL load; islanded operation with balanced RL load to unbalanced RL load; and, finally, islanded operation with unbalanced RL load to non-linear load. In all these results, the Slave inverter is the controller so as to inject the total power demanded by the load into the microgrid, while the Master inverter is responsible for synthesizing appropriate grid voltages, especially when in islanded operation.

Analyzing the results presented in Figs. 3.3, 3.4, 3.5 and 3.6, we can see a very fast transition from grid-connected to islanded mode, depicted in greater detail in Fig. 3.4. However, as the FCS-MPC implements only one voltage vector per switching period T_s, some distortions can be seen in the grid-side currents of the Slave inverter. Also, due to the same reason, there is a steady-state error between the current references and the actual grid-side currents of the Slave inverter. When in grid-connected mode, the remaining current associated to this error is supplied to the load by the grid; on the other hand, when islanded, the Master inverter is responsible for providing these extra currents in order to feed the load with proper active and reactive powers. This is depicted in Fig. 3.6 for the case of the non-linear load, where the Master currents in islanded mode can be seen, as well as its active and reactive powers.

In islanded mode, the very fast transient response of the Slave inverter is seen for all load step changes and variations, including the non-linear and unbalanced loads in Figs. 3.5 and 3.6. The Master inverter also has a very good response, synthesizing proper voltages for the islanded microgrid (apart from minor distortions that appear with the non-linear load). Also, it can be seen that the Slave inverter injects proper

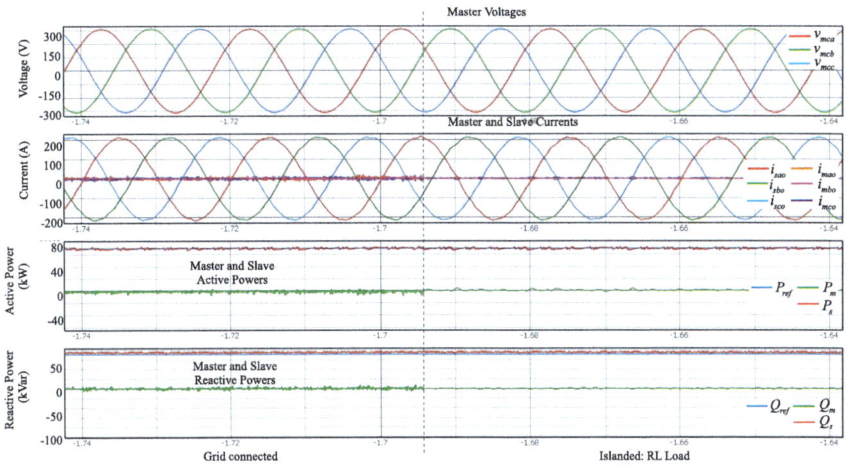

Fig. 3.4 Zoom in the transient response from grid-connected to islanded mode with balanced RL load for the FCS-MPC: (i) capacitor voltages of the LC filter of the Master inverter; (ii) grid-side currents of the Master and Slave inverters; (iii) total active power reference, and Master and Slave active powers; (iv) total reactive power reference, and Master and Slave reactive powers

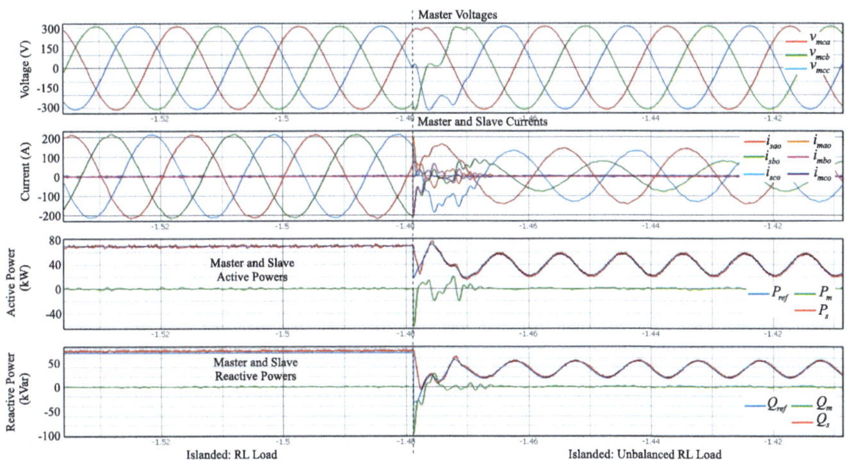

Fig. 3.5 Zoom in the transient response from islanded mode with balanced RL load to islanded mode with unbalanced RL load for the FCS-MPC for the FCS-MPC: (i) capacitor voltages of the LC filter of the Master inverter; (ii) grid-side currents of the Master and Slave inverters; (iii) total active power reference, and Master and Slave active powers; (iv) total reactive power reference, and Master and Slave reactive powers

active and reactive powers, as demanded by the load, while the active and reactive powers of the Master inverter remain close to zero, as its main objective in this case is to synthesize appropriate grid voltages for the microgrid; only with non-linear load the Master supplies current, due to the facts previously discussed. Finally, it must

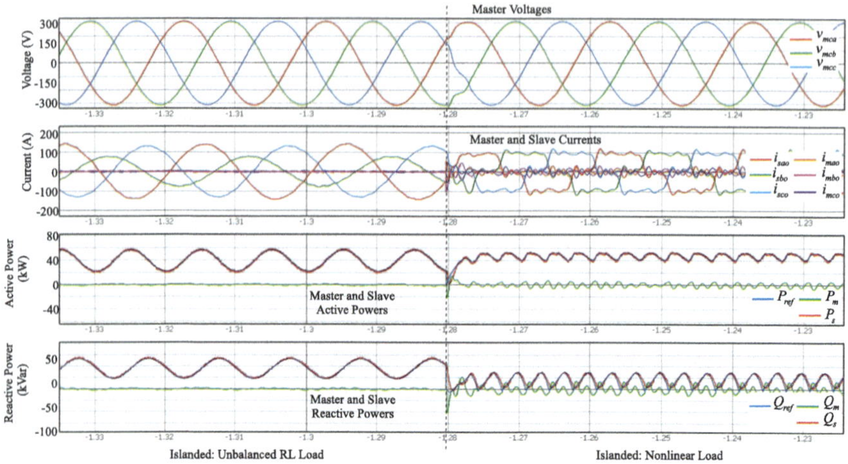

Fig. 3.6 Zoom in the transient response from islanded mode with unbalanced RL load to islanded mode with non-linear load: (i) capacitor voltages of the LC filter of the Master inverter; (ii) grid-side currents of the Master and Slave inverters; (iii) total active power reference, and Master and Slave active powers; (iv) total reactive power reference, and Master and Slave reactive powers

be stressed that the loads and the line and grid impedances are unknown parameters for both the Master and Slave inverters. Consequently, these results show the very good performance of the proposed Master-Slave FSF-MPC controller with respect to uncertainties and parametric variations in the system.

3.4 Proposed Fixed Switching Frequency MPC (FSF-MPC) for Master-Slave Inverters

As previously mentioned, even though FCS-MPC has a very good dynamic and transient performance, it inherently results in variable switching frequency for the Master and Slave inverters, increasing the harmonic content of the output voltages and currents and making the design of the output filter more complex. In order to overcome these issues, a Fixed Switching Frequency Model Predictive Control (FSF-MPC) will now be proposed for the same microgrid shown in Fig. 3.1. The block diagram of the proposed Master-Slave FSF-MPC is shown in Fig. 3.7. In the following sections, the FSF-MPC for both the Master and Slave inverters will be presented.

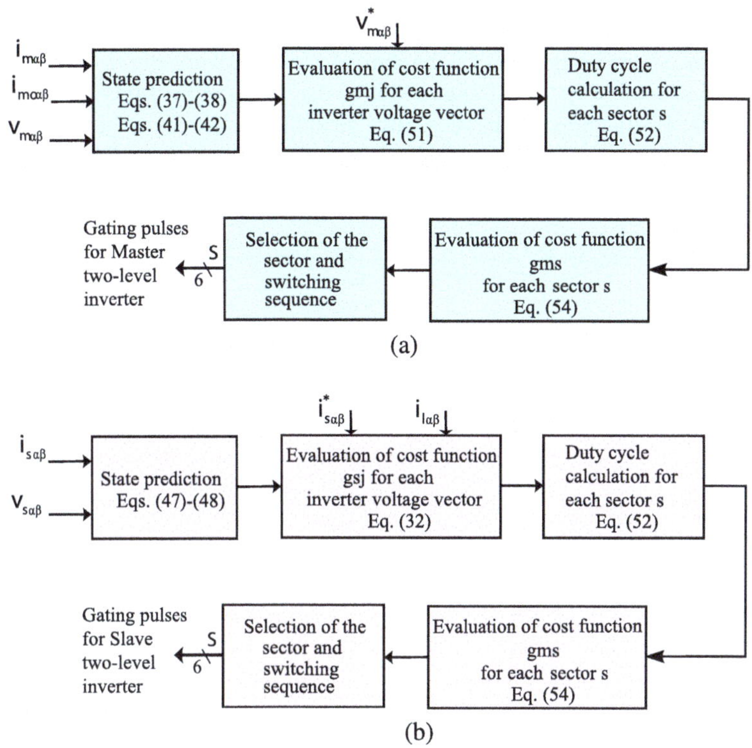

Fig. 3.7 Block diagram of the proposed Master-Slave FSF-MPC: **a** proposed FSF-MPC for Master inverter; **b** proposed FSF-MPC for Slave inverter

3.4.1 Voltage FSF-MPC for the Master Inverter

Let us begin by describing the FSF-MPC for the Master inverter. The same model considerations described previously are also applied here for the FSF-MPC and the discrete-time system equations are equal to (10)–(11) and (18)–(19). However, now the voltage synthesized by the Master inverter during each T_s is not implemented as the inverter voltage vector that minimizes the voltage cost function. Instead, a switching sequence is now implemented over T_s, resulting in fixed switching frequency at the output.

In order to implement the Voltage FSF-MPC, for each T_s, $v_{m\alpha\beta}(k+1)$ is predicted for all inverter voltage vectors, considering the same cost function as for the Voltage FCS-MPC, that is reproduced here again for the sake of clarity:

$$g_{mj} = \left[(v^*_{m\alpha}(k+1) - v_{m\alpha j}(k+1))^2 + (v^*_{m\beta}(k+1) - v_{m\beta j}(k+1))^2\right] \quad (3.51)$$

where $j = 1 \ldots 8$ represents each inverter voltage vector. For each of the six sectors S1 to S6 of the inverter SV diagram, a switching sequence is defined offline, comprised of two non-null voltage vectors and both redundancies of the null vector. Figure 3.8 shows the normalized SV diagram of a two-level inverter in the $\alpha\beta$ coordinates and six sectors S1 to S6. The next step in the Voltage FSF-MPC is to select the sector of the SV diagram and its respective switching sequence to be implemented by the inverter over T_s. To accomplish this, first, the duty cycles of the three vectors of each sector are calculated as [45]

$$d_1 = \frac{g_{m2}g_{m3}}{G} \quad d_2 = \frac{g_{m1}g_{m3}}{G} \quad d_3 = \frac{g_{m1}g_{m2}}{G} \tag{3.52}$$

$$G = (g_{m2}g_{m3} + g_{m1}g_{m3} + g_{m2}g_{m1}) \tag{3.53}$$

where $d_1 + d_2 + d_3 = 1$ and the notation 1, 2 and 3 represents the three vectors of each sector, as depicted in Fig. 3.8.

Based on (3.51)–(3.53), the cost function of each one of the sectors of the SV diagram is

$$g_{ms} = d_1 g_{m1} + d_2 g_{m2} + d_3 g_{m3} \tag{3.54}$$

where $s = 1 \ldots 6$ represents each of the six sectors of the inverter SV diagram. Finally, the sector with the lowest value of its associated cost function is selected and its respective switching sequence is then implemented by the Master inverter over T_s. This procedure is depicted in the upper part of Fig. 3.11.

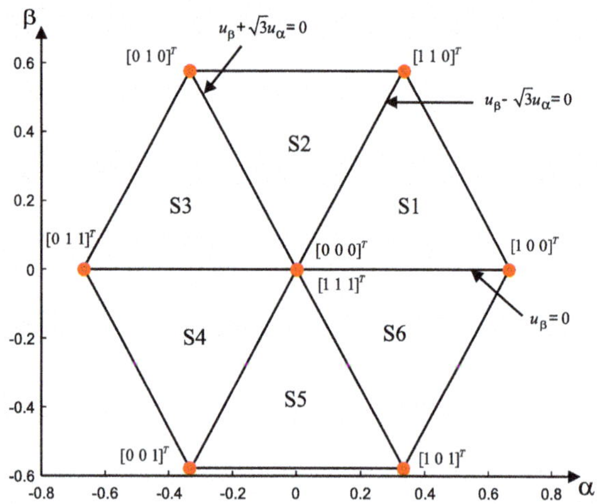

Fig. 3.8 Normalized SV diagram of a two-level inverter in the $\alpha\beta$ coordinates

3.4.2 *Slave Inverter Current MPC*

For the Slave inverter, the procedure for the Current FSF-MPC is the same as for the Master one, and the same model considerations for the Slave inverter described in the previous section are also applied here. The discrete-time system equations are equal to (3.10)–(3.11). In the same way as described for the Master inverter, a switching sequence is defined offline for each of the six sectors of the inverter SV, comprised of two non-null voltage vectors and both redundancies of the null vector. Then, the sector that minimizes the FSF-MPC cost function and its respective switching sequence are selected and implemented by the inverter over T_s. The FSF-MPC cost function for the Slave is repeated here for the sake of clarity:

$$g_{sj} = [(i^*_{s\alpha}(k+1) - i_{s\alpha j}(k+1))^2 + (i^*_{s\beta}(k+1) - i_{s\beta j}(k+1))^2] \tag{3.55}$$

Then, the duty cycles of the three vectors of each sector are calculated as [45]

$$d_1 = \frac{g_{s2}g_{s3}}{G} \quad d_2 = \frac{g_{s1}g_{s3}}{G} \quad d_3 = \frac{g_{s1}g_{s2}}{G} \tag{3.56}$$

$$G = (g_{s2}g_{s3} + g_{s1}g_{s3} + g_{s2}g_{s1}) \tag{3.57}$$

where $d_1 + d_2 + d_3 = 1$. Based on 3.55–3.57, the cost function of each one of the sectors of the SV diagram is

$$g_{ss} = d_1 g_{s1} + d_2 g_{s2} + d_3 g_{s3} \tag{3.58}$$

where $s = 1 \ldots 6$ represents each of the six sectors of the inverter SV diagram and the notation 1, 2 and 3 represent the three vectors of each sector, as depicted in Fig. 3.8. Finally, the sector with the lowest value of its associated cost function is selected and its respective switching sequence is then implemented by the Slave over T_s. This procedure is depicted in the lower part of Fig. 3.11.

3.5 Case Studies for the FSF-MPC

To validate the proposed Master-Slave FSF-MPC, the microgrid of Fig. 3.1 was implemented in the same Typhoon HIL 604 Hardware-in-the-Loop platform used to validate the Master-Slave FCS-MPC. The system parameters are shown in Table 3.1. The Voltage and Current MPCs, as well as other algorithms such as the SRF-PLL, were implemented in the C Block of Typhoon HIL, which includes the processor implementation delay. As a result, the cost functions are calculated for the predicted variables at $k + 2$. The disturbances and the references are also estimated for $k + 2$, as described in (3.37)–(3.50).

The same case studies were implemented for the Master-Slave FSF-MPC and will be described here again. In the beginning, the microgrid is connected to the grid with a balanced three-phase RL load with $R_l = 1.03\,\Omega$ and $L_l = 3.33\,\text{mH}$. Then, the switch that connects it to the main grid is opened, and the microgrid changes to an islanded mode of operation. In islanded mode, the RL load is first changed to an unbalanced three-phase RL load with $R_l = 1.03\,\Omega$ and $L_l = 3.33\,\text{mH}$ for phase a, $R_l = 4\,\Omega$ and $L_l = 10\,\text{mH}$ for phase b and $R_l = 2\,\Omega$ and $L_l = 6\,\text{mH}$ for phase c. Finally, the load is changed to a non-linear load composed of a three-phase diode rectifier with an input L filter with $L = 1\,\text{mH}$ and an output RL load with $R = 5\,\Omega$ and $L = 5\,\text{mH}$. These results are shown in Fig. 3.9, where we can see, from top to bottom, in *abc* coordinates: (i) capacitor voltages of the LC filter of the Master inverter; (ii) grid-side currents of the Master and Slave inverters; (iii) total active power reference (coming from the load demand), and Master and Slave active powers; (iv) total reactive power reference (coming from the load demand), and Master and Slave reactive powers. Also, Figs. 3.10, 3.10 and 3.12 show zooms in the transient results for, respectively, grid-connected to islanded operation with balanced RL load; islanded operation with balanced RL load to unbalanced RL load; and, finally, islanded operation with unbalanced RL load to non-linear load. In all these results, the Slave inverter is controlled so as to inject the total power demanded by the load into the microgrid, while the Master inverter is responsible for synthesizing appropriate grid voltages, especially when in islanded operation.

Looking at the results presented in Figs. 3.9, 3.10, 3.11 and 3.12, we can see a very fast transition from grid-connected to islanded mode, depicted in greater detail in Fig. 3.10. In islanded mode, the very fast transient response of the Slave inverter is seen for all load step changes and variations, including the non-linear and unbalanced loads in Figs. 3.11 and 3.12. The Master inverter also has a very good response, synthesizing proper voltages for the islanded microgrid (apart from minor

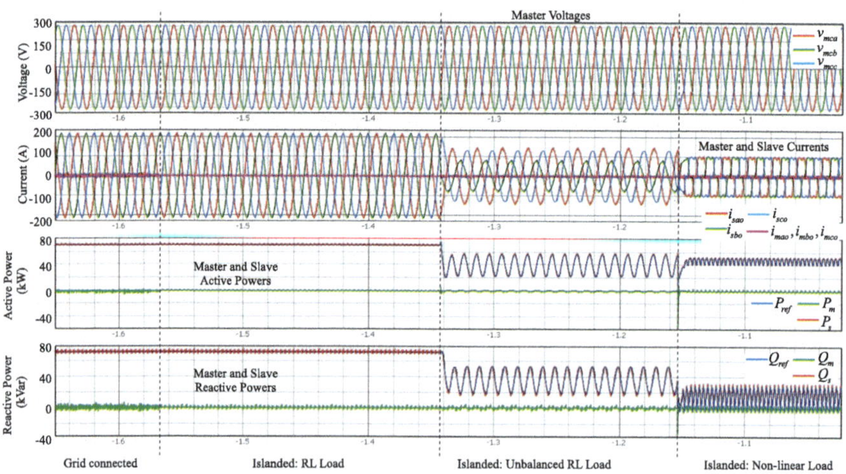

Fig. 3.9 HIL results for the FSF-MPC: (i) capacitor voltages of the LC filter of the Master inverter; (ii) grid-side currents of the Master and Slave inverters; (iii) total active power reference, and Master and Slave active powers; (iv) total reactive power reference, and Master and Slave reactive powers

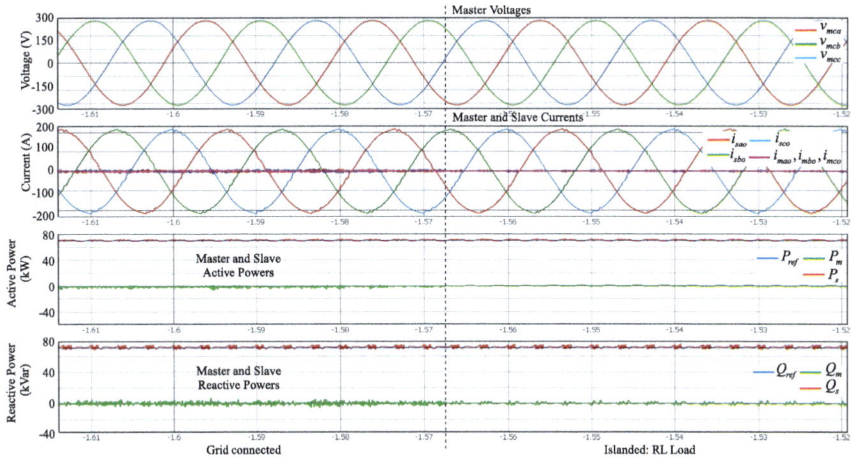

Fig. 3.10 Zoom in the transient response from grid-connected to islanded mode with balanced RL load for the FSF-MPC: (i) capacitor voltages of the LC filter of the Master inverter; (ii) grid-side currents of the Master and Slave inverters; (iii) total active power reference, and Master and Slave active powers; (iv) total reactive power reference, and Master and Slave reactive powers

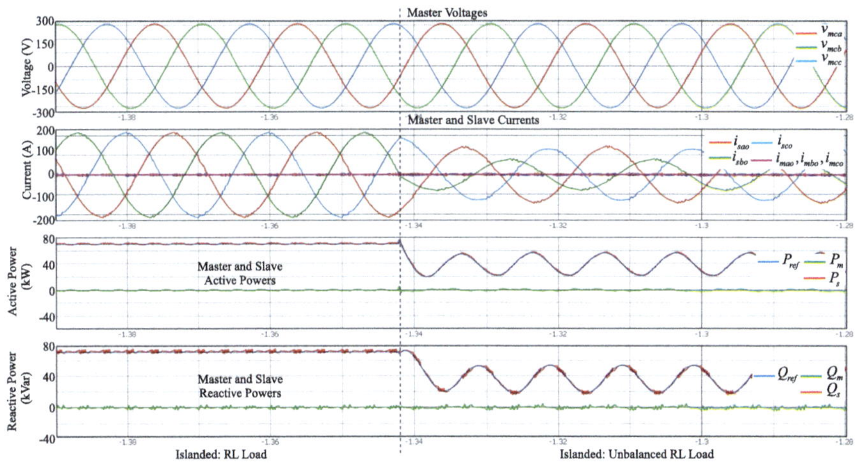

Fig. 3.11 Zoom in the transient response from islanded mode with balanced RL load to islanded mode with unbalanced RL load for the FSF-MPC: (i) capacitor voltages of the LC filter of the Master inverter; (ii) grid-side currents of the Master and Slave inverters; (iii) total active power reference, and Master and Slave active powers; (iv) total reactive power reference, and Master and Slave reactive powers

distortions that appear with the non-linear load). As now both inverters are operating with fixed switching frequency, the steady-state error on currents of the Slave inverter is lower than for the FCS-MPC. This results in less ripple in the output currents. Also, it can be seen that the Slave inverter injects proper active and reactive powers into

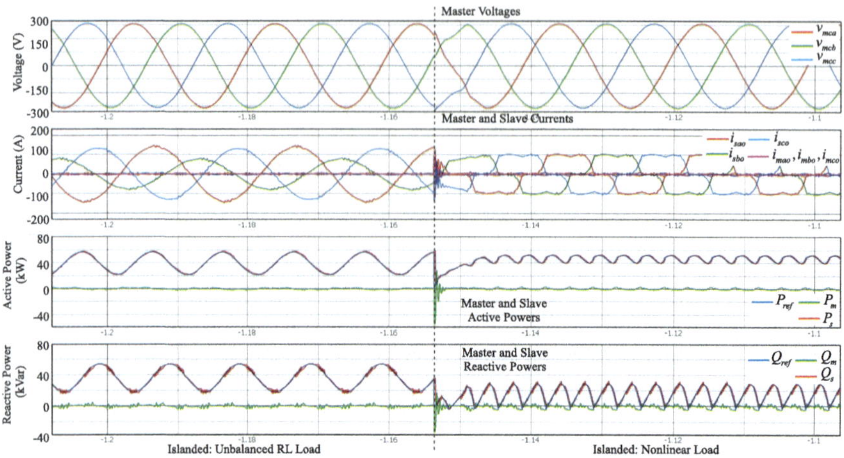

Fig. 3.12 Zoom in the transient response from islanded mode with unbalanced RL load to islanded mode with non-linear load for the FSF-MPC: (i) capacitor voltages of the LC filter of the Master inverter; (ii) grid-side currents of the Master and Slave inverters; (iii) total active power reference, and Master and Slave active powers; (iv) total reactive power reference, and Master and Slave reactive powers

the load, while the active and reactive powers of the Master inverter remain close to zero, even for the non-linear load. For the same case with non-linear load, due to the larger steady-state error inherent to the classical FCS-MPC implementations, the Master had to supply a part of the load currents and power (specially reactive power), as seen in Figs. 3.3 and 3.6.

Another point that is important to mention is that, due to the fixed switching frequency for the FSF-MPC, the parameters of the LCL filter for the Slave inverter are drastically reduced when compared to the FCS-MPC, as seen in Tables 3.1 and 3.2. The values of the filter components for the FCS-MPC were larger than for the FSF-MPC in order to provide low ripples for the grid-side currents. Finally, it must be stressed here again that the load and the line and grid impedances are unknown parameters for both the Master and Slave inverters. Consequently, these results show the very good performance of the proposed Master-Slave FSF-MPC controller with respect to uncertainties and parametric variations in the system.

3.6 Comparative Analysis

In this section, the results presented for the FCS-MPS and the FSF-MPC will be compared with more details. First, let us discuss again about the parameters of the LCL filter of the Slave inverter. For the FCS-MPC, the inverter-side inductor has an inductance of 3mH, the grid-side inductor of 200 μH, and the filter capacitance

Table 3.2 System parameters for the case studies.

Parameter	Value
Total DC-link voltage (Master and Slave)	800 V
Rated power (Master and Slave)	100 kVA
LCL filter (Slave)	500 μH; 100 μF; 50 μH
LC filter (Master)	3 mH; 300 μF
Filter parasitic resistance (Master and Slave)	0.012 Ω
Filter damping resistor Master	1 mΩ
Filter damping resistor Slave	0.2 Ω
Grid impedance	$Z_g = 2.7\,\mu$H, 0.01 Ω
Line impedance	$Z_l = 20\,\mu$H, 0.01 Ω
Grid voltage	220 Vrms; 50 Hz
Sampling frequency f_s	20 kHz
Switching frequency f_s	10 kHz

is equal to 300 μF. On the other hand, for the FSF-MPC, the same parameters are, respectively, 500 μH, 50μH and 100 μF. The values of the filter components for the FCS-MPC are larger than the ones for the FSF-MPC in order to provide low ripples for the grid-side currents, as the FCS-MPC operates with variable switching frequency. To have a less bulky filter, the FCS-MPC would have to be implemented with a larger sampling frequency. To visualize these differences, Figs. 3.13 and 3.14 show, respectively, the grid-side currents of the Slave inverter and zoom of its corresponding spectra with FCS-MPC and the LCL filter parameters equal to the ones for the FSF-MPC. It can be observed that the three currents have very large ripples, resulting in a widespread harmonic content specially in the low-frequency range.

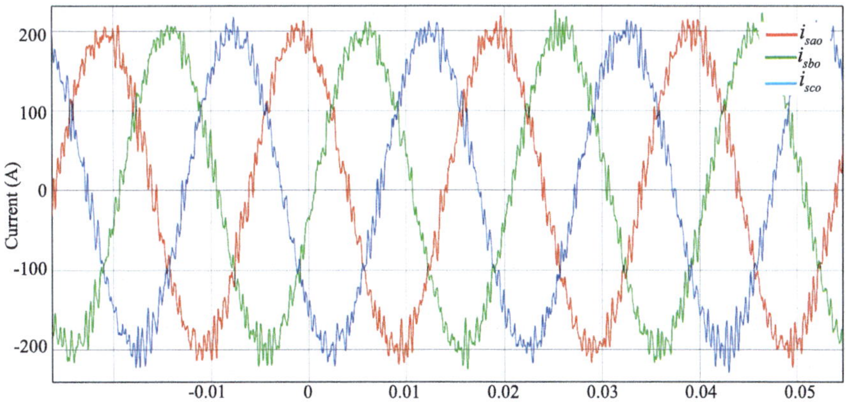

Fig. 3.13 Grid-side currents of the Slave inverter with FCS-MPC and LCL filter parameters equal to the ones for the FSF-MPC

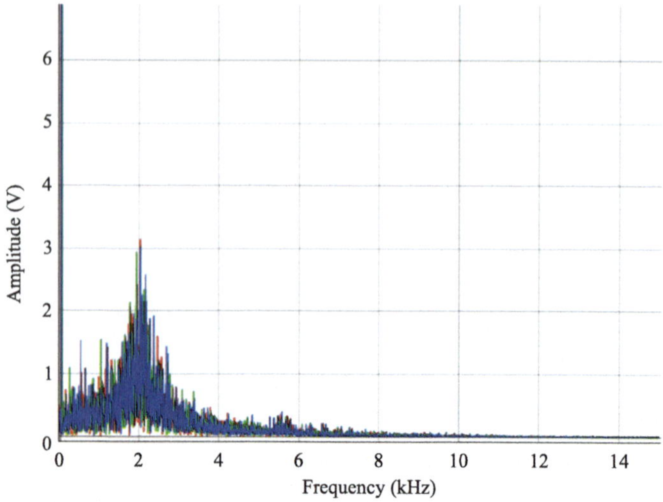

Fig. 3.14 FFT of the grid-side currents of the Slave inverter with FCS-MPC and LCL filter parameters equal to the ones for the FSF-MPC

On the other hand, Figs. 3.15 and 3.16 show, respectively, the grid-side currents of the Slave inverter and a zoom of its corresponding spectra with FCS-MPC and the LCL filter parameters of Table 3.1. It can be seen that the currents in Fig. 3.15 have much lower ripples, and the amplitude of the harmonic components shown in the FFT of Fig. 3.16 are also smaller than for Fig. 3.14. These results, however, are obtained at the expanse of a bulky output LCL filter.

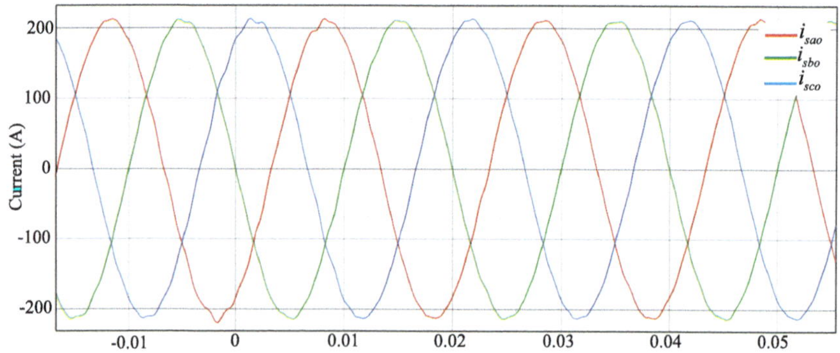

Fig. 3.15 Grid-side currents of the Slave inverter with FCS-MPC and LCL filter parameters equal to the ones shown in Table 3.1

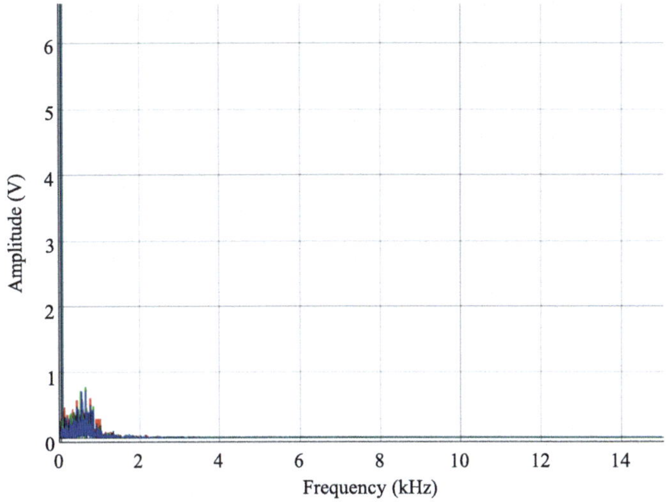

Fig. 3.16 FFT of the grid-side currents of the Slave inverter with FCS-MPC and LCL filter parameters equal to the ones shown in Table 3.1

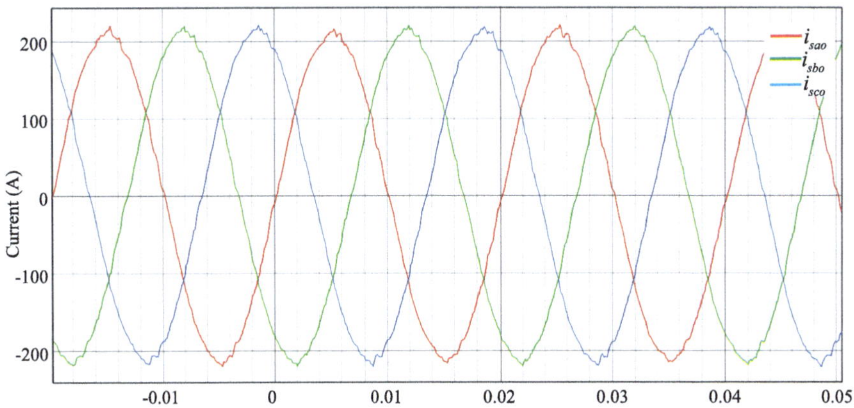

Fig. 3.17 Grid-side currents of the Slave inverter with FSF-MPC and LCL filter parameters equal to the ones shown in Table 3.2

Finally, the grid-side currents of the Slave inverter with FSF-MPC and LCL filter parameters as presented in Table 3.2 are shown in Fig. 3.17, while the zoom on the FFT of the currents is depicted in Fig. 3.18. As expected, the grid-side currents have very low ripple. However, when observing the FFTs in Fig. 3.18, we can notice that the low-order harmonics are bigger than the ones for the FCS-MPC shown in Fig. 3.16. This is due to the fact that the parameters of the LCL filter for the results presented in Fig. 3.16 are very large. The FSF-MPC can be improved by redesigning the LCL filter and/or implementing different switching sequences for the Slave inverter.

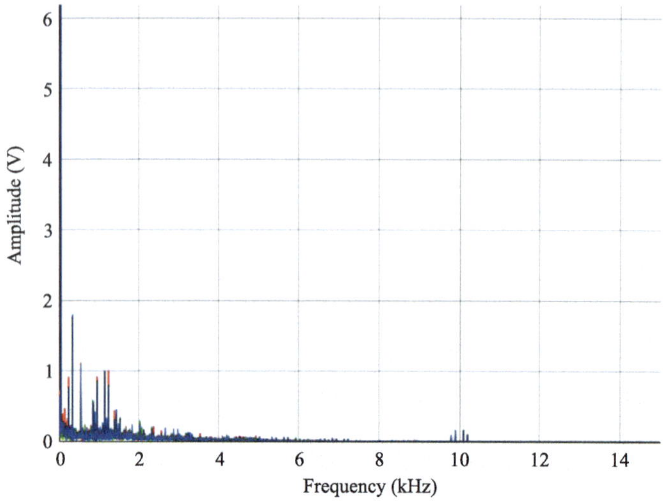

Fig. 3.18 FFT of the grid-side currents of the Slave inverter with FSF-MPC and LCL filter parameters equal to the ones shown in Table 3.2

Regarding the switching frequency analysis, the variable switching frequency of the FCS-MPC is shown in Fig. 3.19, that presents the FFT of the output line-to-line voltages v_{ab} of the Master and the Slave inverters. The fixed switching frequency of the proposed FSF-MPC is shown in Fig. 3.20, that also presents the FFT of the output line-to-line voltages v_{ab} of the Master and the Slave inverters. It can be seen that, for the FSF-MPC, the first harmonics appear at 10kHz for both inverters, validating the operation with fixed switching frequency, while for the FCS-MPC, the harmonic spectra of both voltages are widespread. Also, the low-order harmonics of the FSF-MPPC have very low amplitudes, when compared to the spectra of the output line-to-line voltages obtained with the FCS-MPC. Moreover, irrespective of these differences, both the Master-Slave implementations with FCS-MPC and FSF-MPC presented very good performances, with fast dynamic responses, possibility of multi-variable control, and robustness to parametric uncertainties and variations.

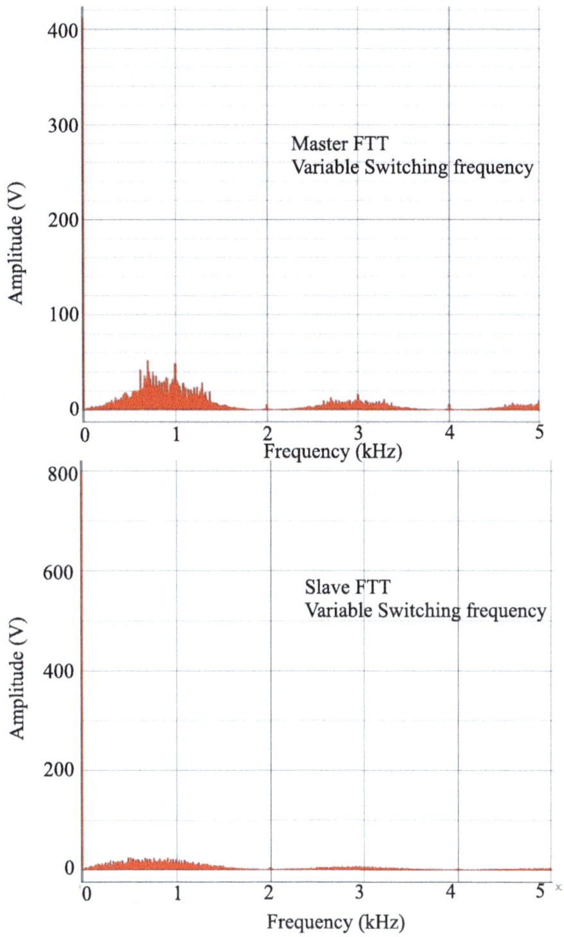

Fig. 3.19 FTT of the output line-to-line voltages v_{ab} of the Master and Slave inverters for the FCS-MPC

3.7 Conclusions

This book chapter presented two Model Predictive Control (MPC) strategies for Master-Slave parallel inverters in microgrids. In this application, the Master is a grid-forming inverter with LC output filter, while the Slave is a grid-following inverter with LCL filter. The microgrid is also composed of linear and non-linear loads and the grid and line impedances. First, a Finite Control Set MPC (FCS-MPC) was presented and validated with Hardware-in-the-Loop (HIL) results for various operational conditions of the microgrid, such as grid connection, islanded mode, and load variations. Afterward, a Fixed Switching Frequency Model Predictive Control (FSF-MPC) was presented for the same microgrid and was also validated through HIL for the same

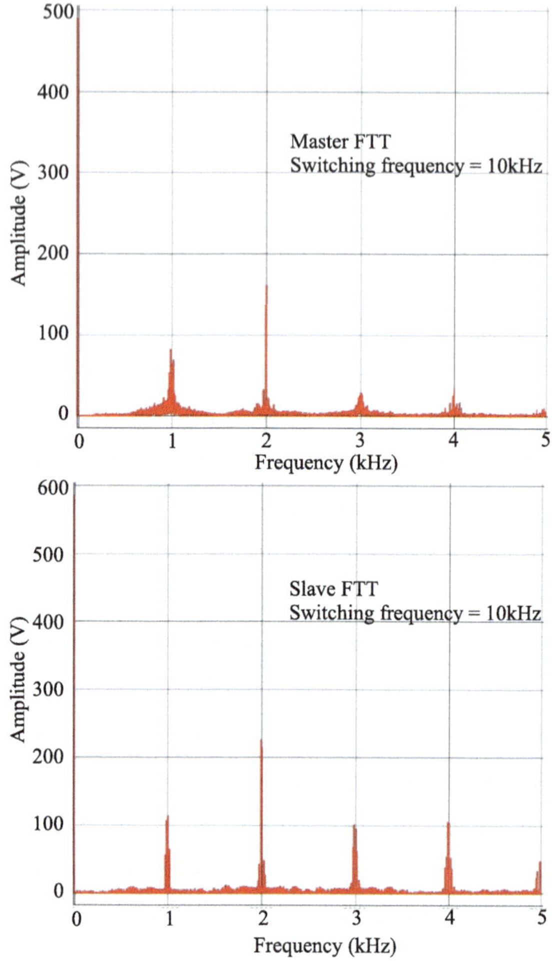

Fig. 3.20 FTT of the output line-to-line voltages v_{ab} of the Master and Slave inverters for the FSF-MPC

operational conditions as for the FCS-MPC. Finally, the performance of both Master-Slave MPCs was compared, demonstrating the good performances of the proposed FCS-MPC and FSF-MPC approaches, such as fast dynamic response, multi-variable control, and robustness to parametric uncertainties.

Acknowledgements This work was supported by ANID, Chile FONDECYT Iniciacion 11230430, SERC-Chile ANID/FONDAP/1522A0006, AC3E (ANID/BASAL/FB0008), FONDECYT 1230250, and ANID projects 1210208 and 1221293.

References

1. Z. Zhang, O. Babayomi, T. Dragicevic, R. Heydari, C. Garcia, J. Rodriguez, R. Kennel, Advances and opportunities in the model predictive control of microgrids: Part i-primary layer. Int. J. Electr. Power Energy Syst. **134**, 107411 (2022)
2. J. Hu, Y. Shan, J.M. Guerrero, A. Ioinovici, K.W. Chan, J. Rodriguez, Model predictive control of microgrids—an overview. Renew. Sustain. Energy Rev. **136**, 110422 (2021)
3. Y. Khayat, Q. Shafiee, R. Heydari, M. Naderi, T. Dragicevic, J.W. Simpson-Porco, F. Dorfler, M. Fathi, F. Blaabjerg, J.M. Guerrero, H. Bevrani, On the secondary control architectures of AC microgrids: an overview. IEEE Trans. Power Electron. **35**(6), 6482 (2020). https://doi.org/10.1109/tpel.2019.2951694
4. A. Alfergani, K.A. Alfaitori, A. Khalil, N. Buaossa, Control strategies in ac microgrid: a brief review, in *2018 9th International Renewable Energy Congress (IREC)* (2018), pp. 1–6. https://doi.org/10.1109/IREC.2018.8362575
5. O. Babayomi, Z. Zhang, T. Dragicevic, R. Heydari, Y. Li, C. Garcia, J. Rodriguez, R. Kennel, Advances and opportunities in the model predictive control of microgrids: Part ii-secondary and tertiary layers. Int. J. Electr. Power Energy Syst. **134**, 107339 (2022)
6. J.G.J. Hu, S. Islam, *Model Predictive Control For Microgrids—From Power Electronics Converters to Energy Management* (IET—The Institution of Engineering and Technology, 2021)
7. Z. Chen, K. Wang, Z. Li, T. Zheng, A review on control strategies of ac/dc micro grid, in *2017 IEEE International Conference on Environment and Electrical Engineering and 2017 IEEE Industrial and Commercial Power Systems Europe (EEEIC/I CPS Europe)* (2017), pp. 1–6. https://doi.org/10.1109/EEEIC.2017.7977807
8. M. Prodanovic, T. Green, H. Mansir, A survey of control methods for three-phase inverters in parallel connection, in *2000 Eighth International Conference on Power Electronics and Variable Speed Drives (IEE Conf. Publ. No. 475)* (2000), pp. 472–477. https://doi.org/10.1049/cp:20000293
9. J. Hu, Y. Shan, K.W. Cheng, S. Islam, Overview of power converter control in microgrids—challenges, advances, and future trends. IEEE Trans. Power Electron. **37**(8), 9907 (2022). https://doi.org/10.1109/tpel.2022.3159828
10. O. Babayomi, Y. Li, Z. Zhang, R. Kennel, J. Kang, Overview of model predictive control of converters for islanded AC microgrids, in *2020 IEEE 9th International Power Electronics and Motion Control Conference (IPEMC2020-ECCE Asia)* (IEEE, 2020). https://doi.org/10.1109/ipemc-ecceasia48364.2020.9368069
11. I. Khan, A.S. Vijay, S. Doolla, Nonlinear load harmonic mitigation strategies in microgrids: state of the art. IEEE Syst. J. **16**(3), 4243 (2022). https://doi.org/10.1109/jsyst.2021.3130612
12. P. Buduma, M.K. Das, R.T. Naayagi, S. Mishra, G. Panda, Seamless operation of master-slave organized AC microgrid with robust control, islanding detection, and grid synchronization. IEEE Trans. Indus. Appl. **58**(5), 6724 (2022). https://doi.org/10.1109/tia.2022.3185575
13. A. Mortezaei, M.G. Simoes, M. Savaghebi, J.M. Guerrero, A. Al-Durra, Cooperative control of multi-master-slave islanded microgrid with power quality enhancement based on conservative power theory. IEEE Trans. Smart Grid **9**(4), 2964 (2018). https://doi.org/10.1109/tsg.2016.2623673
14. D. Miller, G. Mirzaeva, C.D. Townsend, G.C. Goodwin, Decentralised droopless control of islanded radial AC microgrids without explicit communication. IEEE Open J. Indus. Appl. **3**, 104 (2022). https://doi.org/10.1109/ojia.2022.3177857
15. Y. Yang, R.J. Wai, Self-constructing fuzzy-neural-network-imitating sliding-mode control for parallel-inverter system in grid-connected microgrid. IEEE Access **9**, 167389 (2021). https://doi.org/10.1109/access.2021.3135856
16. D.D. Zheng, S.S. Madani, A. Karimi, Data-driven distributed online learning control for islanded microgrids. IEEE J. Emerg. Selected Top. Circ. Syst. **12**(1), 194 (2022). https://doi.org/10.1109/jetcas.2022.3152938

17. J. Rodriguez, M.P. Kazmierkowski, J.R. Espinoza, P. Zanchetta, H. Abu-Rub, H.A. Young, C.A. Rojas, State of the art of finite control set model predictive control in power electronics. IEEE Trans. Ind. Inform. **9**(2), 1003 (2013)
18. S. Vazquez, J. Rodriguez, M. Rivera, L.G. Franquelo, M. Norambuena, Model predictive control for power converters and drives: advances and trends. IEEE Trans. Ind. Electron. **64**(2), 935 (2017). https://doi.org/10.1109/TIE.2016.2625238
19. P. Karamanakos, E. Liegmann, T. Geyer, R. Kennel, Model predictive control of power electronic systems: methods, results, and challenges. IEEE Open J. Ind. Appl. **1**, 95 (2020). https://doi.org/10.1109/OJIA.2020.3020184
20. M. Aly, F. Carnielutti, J. Rodriguez, M. Norambuena, S. Kouro, A.K. Rathore, Finite control set model predictive control without weighting factors for common grounded five-level PV inverter, in *2021 IEEE 30th International Symposium on Industrial Electronics (ISIE)* (2021), pp. 01–06. https://doi.org/10.1109/ISIE45552.2021.9576284
21. C.R.D. Osório, D.A. Schuetz, G.G. Koch, F. Carnielutti, D.M. Lima, L.A.M. Jr, V.F. Montagner, H. Pinheiro, Modulated model predictive control applied to lcl-filtered grid-tied inverters: a convex optimization approach. IEEE Open J. Indus. Appl. **2**, 366 (2021). https://doi.org/10.1109/OJIA.2021.3134585
22. S.R. Mohapatra, V. Agarwal, Model predictive controller with reduced complexity for grid-tied multilevel inverters. IEEE Trans. Indus. Electron. **66**(11), 8851 (2019). https://doi.org/10.1109/TIE.2018.2866115
23. D. Xiao, K.S. Alam, M. Norambuena, M.F. Rahman, J. Rodriguez, Modified modulated model predictive control strategy for a grid-connected converter. IEEE Trans. Indus. Electron. **68**(1), 575 (2021). https://doi.org/10.1109/TIE.2020.2965457
24. X. Chen, W. Wu, N. Gao, H.S.H. Chung, M. Liserre, F. Blaabjerg, Finite control set model predictive control for lcl-filtered grid-tied inverter with minimum sensors. IEEE Trans. Indus. Electron. **67**(12), 9980 (2020). https://doi.org/10.1109/TIE.2019.2962444
25. J. Scoltock, T. Geyer, U.K. Madawala, Model predictive direct power control for grid-connected NPC converters. IEEE Trans. Indus. Electron. **62**(9), 5319 (2015). https://doi.org/10.1109/TIE.2015.2410259
26. M.G. Judewicz, S.A. González, J.R. Fischer, J.F. Martínez, D.O. Carrica, Inverter-side current control of grid-connected voltage source inverters with lcl filter based on generalized predictive control. IEEE J. Emerg. Selected Top. Power Electron. **6**(4), 1732 (2018). https://doi.org/10.1109/JESTPE.2018.2826365
27. M. Aly, F. Carnielutti, A. Shawky, E.M. Ahmed, M. Norambuena, S. Kouro, J. Rodriguez, Weighting factorless sequential model predictive control method with fixed switching frequency for five-level t-type photovoltaic inverters, in *IECON 2021—47th Annual Conference of the IEEE Industrial Electronics Society* (2021), pp. 1–6. https://doi.org/10.1109/IECON48115.2021.9589758
28. Y. Shan, J. Hu, Z. Li, J.M. Guerrero, A model predictive control for renewable energy based ac microgrids without any PID regulators. IEEE Trans. Power Electron. **33**(11), 9122 (2018). https://doi.org/10.1109/TPEL.2018.2822314
29. T. Dragičević, Model predictive control of power converters for robust and fast operation of ac microgrids. IEEE Trans. Power Electron. **33**(7), 6304 (2018). https://doi.org/10.1109/TPEL.2017.2744986
30. A.B. Kunya, M. Argin, Model predictive load frequency control of multi-area interconnected power system, in *2018 IEEE Texas Power and Energy Conference (TPEC)* (2018), pp. 1–6. https://doi.org/10.1109/TPEC.2018.8312114
31. A. Venkat, I. Hiskens, J. Rawlings, S. Wright, Distributed MPC strategies with application to power system automatic generation control. IEEE Trans. Control Syst. Technol. **16**(6), 1192 (2008). https://doi.org/10.1109/tcst.2008.919414
32. G. Lou, W. Gu, Y. Xu, M. Cheng, W. Liu, Distributed mpc-based secondary voltage control scheme for autonomous droop-controlled microgrids. IEEE Trans. Sustain. Energy **8**(2), 792 (2017). https://doi.org/10.1109/TSTE.2016.2620283

33. C. Ahumada, R. Cárdenas, D. Sáez, J.M. Guerrero, Secondary control strategies for frequency restoration in islanded microgrids with consideration of communication delays. IEEE Trans. Smart Grid **7**(3), 1430 (2016). https://doi.org/10.1109/TSG.2015.2461190
34. J. Sachs, O. Sawodny, A two-stage model predictive control strategy for economic diesel-pv-battery island microgrid operation in rural areas. IEEE Trans. Sustain. Energy **7**(3), 903 (2016). https://doi.org/10.1109/TSTE.2015.2509031
35. A. Parisio, E. Rikos, L. Glielmo, A model predictive control approach to microgrid operation optimization. IEEE Trans. Control Syst. Technol. **22**(5), 1813 (2014). https://doi.org/10.1109/TCST.2013.2295737
36. J. Ospina, N. Gupta, A. Newaz, M. Harper, M.O. Faruque, E.G. Collins, R. Meeker, G. Lofman, Sampling-based model predictive control of pv-integrated energy storage system considering power generation forecast and real-time price. IEEE Power Energy Technol. Syst. J. **6**(4), 195 (2019). https://doi.org/10.1109/JPETS.2019.2935703
37. Y. Shan, J. Hu, K. Wai Cheng, M. Liu, A universal model predictive control for practical ac microgrids with pvs and battery energy storage systems, in *2018 IEEE Energy Conversion Congress and Exposition (ECCE)* (2018), pp. 6257–6262. https://doi.org/10.1109/ECCE.2018.8557588
38. Y. Shan, J. Hu, K.W. Chan, S. Islam, A unified model predictive voltage and current control for microgrids with distributed fuzzy cooperative secondary control. IEEE Trans. Indus. Inform. **17**(12), 8024 (2021). https://doi.org/10.1109/TII.2021.3063282
39. S. Shahzad, M.A. Abbasi, M.A. Chaudhry, M.M. Hussain, Model predictive control strategies in microgrids: a concise revisit. IEEE Access **10**, 122211 (2022). https://doi.org/10.1109/ACCESS.2022.3223298
40. O. Babayomi, Z. Zhang, T. Dragicevic, J. Hu, J. Rodriguez, Smart grid evolution: predictive control of distributed energy resources–a review. Int. J. Electr. Power Energy Syst. **147**, 108812 (2023). https://doi.org/10.1016/j.ijepes.2022.108812
41. Y. Shan, J. Hu, H. Liu, A holistic power management strategy of microgrids based on model predictive control and particle swarm optimization. IEEE Trans. Indus. Inform. **18**(8), 5115 (2022). https://doi.org/10.1109/TII.2021.3123532
42. Y. Shan, J. Hu, H. Liu, Economic dispatch of model predictive controlled distributed power generation, in *2021 IEEE International Conference on Predictive Control of Electrical Drives and Power Electronics (PRECEDE)* (2021), pp. 836–839. https://doi.org/10.1109/PRECEDE51386.2021.9680878
43. E. Hajimalekkheyli, A. Ramezani, S. Amanzadeh, Frequency and voltage rapid restoration in networked microgrids with model predictive controller, in *2022 9th Iranian Conference on Renewable Energy & Distributed Generation (ICREDG)* (2022), pp. 1–9. https://doi.org/10.1109/ICREDG54199.2022.9804566
44. M. Mottaghizadeh, F. Aminifar, T. Amraee, M. Sanaye-Pasand, Robust fuzzy model predictive control for voltage regulation in islanded microgrids. IET Gener. Transm. Distrib. **16**(5), 1013 (2021). https://doi.org/10.1049/gtd2.12345
45. Y. Yang, J. Pan, H. Wen, X. Zhang, M. Norambuena, L. Xu, J. Rodriguez, Computationally efficient model predictive control with fixed switching frequency of five-level anpc converters. IEEE Trans. Ind. Electron. **69**(12), 11903 (2022). https://doi.org/10.1109/TIE.2021.3131797
46. D. Schuetz, C.R.D. Osório, L.A. Maccari, D.M. Lima, F.d.M. Carnielutti, V.F. Montagner, H. Pinheiro, Space vector modulated model predictive control for grid-tied converters. IEEE Trans. Indus. Inform.**19**(1), 414 (2023). https://doi.org/10.1109/TII.2022.3193674
47. F. Carnielutti, M. Aly, M. Norambuena, J. Rodriguez, Model predictive control for master-slave inverters in microgrids, in *IECON 2022—48th Annual Conference of the IEEE Industrial Electronics Society* (2022), pp. 1–6. https://doi.org/10.1109/IECON49645.2022.9968694
48. R. Teodorescu, M. Liserre, P. Rodriguez, *Grid Converters for Photovoltaic and Wind Power Systems* (Wiley-IEEE Press, 2007)

Chapter 4
Distributed Secondary Control of Microgrids with Inertia Emulation

Zhenbin Zhang and Oluleke Babayomi

4.1 Introduction

A microgrid is an electrical network that comprises distributed generation, distribution lines, and consumer loads, with the capability to operate autonomously and can be connected to the grid as well. Sustainable microgrids are powered by renewable distributed energy resources (DER) as shown in Fig. 4.1. The control of microgrids is a hierarchical problem at primary secondary and tertiary control levels. This facilitates the multi-objective, and multi-timescale control requirements of the microgrid as shown in Fig. 4.2. The primary level is characterized by fast changing signals controlled by digital signal processors operating at high sampling rates and timescale of milliseconds. The secondary level operates in seconds and is necessary to compensate for deviations at the primary level. The tertiary level operates in the timescale of hours. These three hierarchies are cascaded, with each outer layer providing additional stability and references for the inner control layers.

Microgrid DERs have their power output conditioned by power electronic converters, resulting in relatively poorer inertia than the main grid that runs mainly on large synchronous generators. For instance, wind turbine systems utilize variable frequency power conversion systems which decouple the rotatory mechanical system from the electrical power system, resulting in weaker inertia. Weak power system inertia makes the electrical network vulnerable to frequency instability due to the

Z. Zhang (✉) · O. Babayomi
School of Electrical Engineering, Shandong University, Jinan 250061, China
e-mail: zbz@sdu.edu.cn

O. Babayomi
e-mail: oluleke.babayomi@mail.sdu.edu.cn

O. Babayomi
CCS Graduate School of Mobility, Korea Advanced Institute of Science and Technology, Daejeon 34141, Republic of Korea

J. Hu (ed.), *Community Energy and Microgrids*, Green Energy and Technology,
https://doi.org/10.1007/978-981-97-6297-2_4

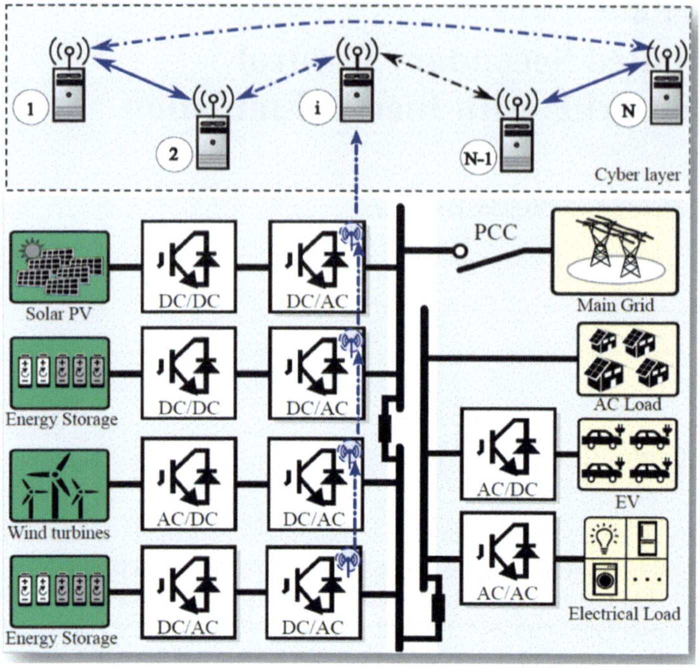

Fig. 4.1 A microgrid powered by renewable energy sources

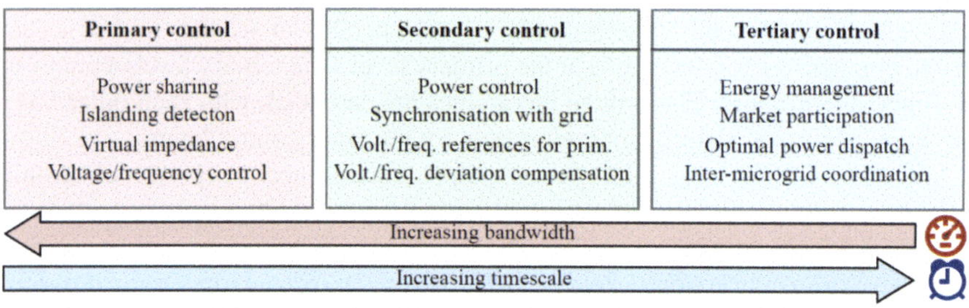

Fig. 4.2 Hierarchical control levels of a microgrid

inability to recover quickly from transients arising from sudden generator disconnections or high load changes. The 2019 British Power Grid blackout is an example of grid failure arising from low inertia due to increasing penetration of distributed energy systems.

The fast frequency regulation of grid-forming voltage source converters (GFVSC) is commonly implemented with droop control, due to its simplicity and maturity of implementation. Nonetheless, droop causes frequency deviations (during load changes) that can be corrected by the hierarchical stages earlier described. At the primary level, frequency response is enhanced by methods including virtual synchronous generators (VSG) [1] and inertia-improving hybrid energy storage installations [2]. These techniques can significantly decrease the frequency drift by reducing the rate of change of frequency (ROCOF).

Further enhancements to frequency control can be achieved by appropriate secondary-level control to compensate for primary-level deviations. Stanojev et al. [3] reported improvements to microgrid frequency nadir and rate of change of frequency (ROCOF) using centralized [4] and decentralized model predictive control (MPC). Although MPC can give a faster dynamic response than linear control, centralized and decentralized secondary control methods are limited in scalability when the DERs increase in number. The receding horizon-based minimization of ROCOF was the subject of frequency of [5]. However, the comprehensive voltage and secondary control were not investigated.

Centralized secondary control methods for GFVSCs are limited in scale and also are vulnerable to failure when the central power converter becomes faulty [4]. On the other hand, decentralized schemes [6] are effective for microgrids with a relatively low number of power converters. The reliability becomes questionable for large-sized microgrids; therefore, distributed control schemes are desirable. Related studies usually employ linear control at the primary level, but this slows the dynamic response of the system. Nonetheless, nonlinear controllers like MPC have been shown to provide fast-dynamic response for GFVSCs [7–9]. A recent study proposed the use of VSG-MPC with faster dynamic response for the primary control and distributed control for frequency and voltage regulation [10]. However, the design of the consensus controller for the secondary hierarchy was done with a high number of sensors and assumed perfect communication conditions. Therefore, the objective of this chapter is the distributed consensus-based control of voltage and frequency in microgrids with GFVSCs with consideration for low communication bandwidth and channel uncertainties.

4.2 Overview of Control Techniques of Distributed Energy Resources in Microgrids

4.2.1 Primary Control Techniques

The primary level of control functions within milliseconds to seconds and regulates the frequency, voltage, and power within constraints that ensure microgrid stability and safety of connected electrical devices. It is important to stress that the primary controller has the first and fastest response to clear transients arising from load and source disturbances. There are usually two levels of primary control, i.e., the low level and high level of control. At the lowest level, voltage and frequency levels are closely monitored to ensure acceptable deviation from nominal values. The higher primary level caters to power sharing, virtual impedance emulation, and islanding detection.

Different primary control methods include centralized, master–slave, droop, droop-free distributed control, virtual oscillator control, and MPC.

Centralized Control

Centralized control operates for primary control of multiple converters and has two control loops: the inner current control loop and an outer voltage control loop. First, the voltage controller regulates the load voltage to the reference. The current controller uses the average of the total load current as a reference to the individual current in each converter. Subsequently, pulse width modulation sends switching signals to the converter. This technique becomes costly as the microgrid grows in size.

Master–Slave Control

The master–slave control method has a dedicated master module inverter that supplies the AC voltage to the load. The other modules in the system also get their current references from the master module [11]. Thus, load power is shared among the converters by the current sharing. An advantage of this method is that it does not require a dedicated phase-locked loop. Furthermore, stability and disturbance rejection are enhanced by voltage (and current) feed-forward control of the slave modules.

Droop Control

Droop control is derived from the traditional droop operation of the governor of a synchronous generator, whereby the angular speed decreases when the load increases and the speed increases when the load decreases. This principle helps power converters to carry out power sharing at low cost, i.e., without communication requirements [12].

Virtual Synchronous Generators

VSGs also operate with the droop principle. The main difference from the earlier discussed droop is that the swing equation of the synchronous generator is emulated. This results in improved transient response due to the inclusion of inertia in the model [10]. Therefore, lower rates of change of frequency can be achieved, compared with the conventional droop method.

Virtual Oscillator Control (VOC)

VOC mimics the dynamics of limit-cycle oscillators for application in microgrids [13]. The VOC has an oscillator comprising passive RLC components and a voltage-dependent current source. Its operation is remarkably different from the preceding droop control methods because the VOC's only input is the instantaneous output current. Therefore, its dynamic response is faster than droop control. The limitation of VOC in not having an effective synchronization to the grid was studied in [14]; an effective solution was presented for both islanded and grid-connected operations. Also, the third-order harmonic usually associated with VOC was attenuated for grid-tied converters in [15].

Other primary control methods include droop-free distributed control (which works with the communication-based cooperative control principle and obviates any need for secondary control), and advanced control methods like sliding mode control and MPC [16].

4.2.2 Secondary Control Techniques

Secondary control of microgrids is usually necessary if droop (or related) control is applied at the primary level—which results in deviations during active and reactive power changes of the load. These secondary techniques, namely centralized, decentralized, and distributed control, operate with(out) dedicated communication links for data sharing.

Centralized secondary control (shown in Fig. 4.3) has one control system that regulates multiple voltage source converters (VSC) in the microgrid network. However, this control technique requires a high bandwidth communication link and is vulnerable to systemic failure if the controller malfunctions. Decentralized control (shown in Fig. 4.4) has multiple secondary controllers and so, is more reliable due to a measure of redundancy; however, its reliability is limited to small-area systems. The distributed control (consensus-based) architecture (shown in Fig. 4.5) is more reliable for wide-area systems, preserves the privacy of agents, and can function well over low-bandwidth systems with data latency. However, it is more complex in design and control.

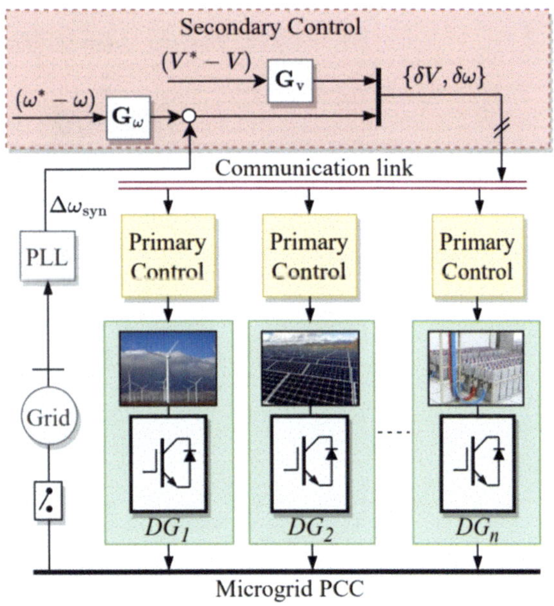

Fig. 4.3 Microgrid centralized secondary control architecture [17]

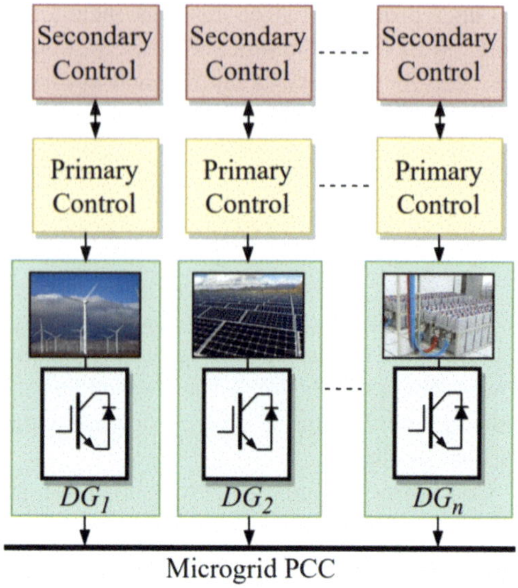

Fig. 4.4 Microgrid decentralized secondary control architecture [17]

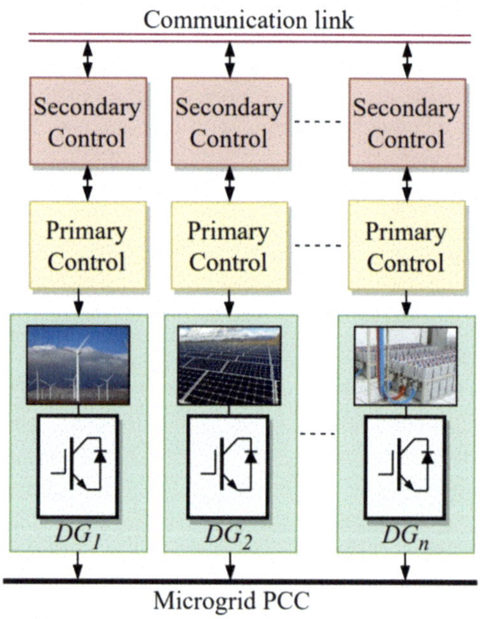

Fig. 4.5 Microgrid distributed secondary control architecture [17]

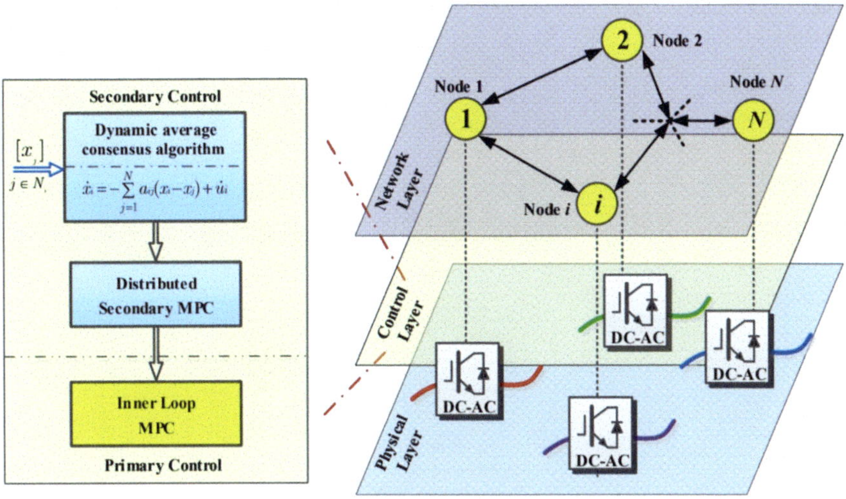

Fig. 4.6 Distributed MPC of microgrid (where a_{ij}, x, and u are the adjacency matrix, state variable, and control input respectively) [17]

Distributed Secondary Control Techniques

The distributed secondary control methods are of three types: averaging schemes, consensus-based, and event-triggered methods. Averaging distributed control has a goal of computing the averaged instantaneous signal (frequency or voltage)—this can be done by different techniques, for instance, consensus-based average observers or other methods. The global average for all modules in the system is then communicated to all the units as references in the secondary control, e.g., the distributed dynamic average consensus method shown in Fig. 4.6. Distributed consensus schemes achieve control share information among neighbor units and utilize the information for system parameter computations.

The controller can operate with sparse bandwidth and is robust to communication link uncertainties including data dropouts and latency. Event-triggered distributed control improves the efficiency of the control scheme by requiring data exchange in the network only when the monitored signals satisfy a predetermined condition. The event-triggered scheme avoids communication link overload and can be time-triggered, event-triggered, and self-triggered sampling, as shown in Fig. 4.7 (see Table 4.1).

4.2.3 Tertiary Control

The global stability of the microgrid is overseen at the tertiary control level. This includes the interaction between the microgrid and main grid which is influenced by

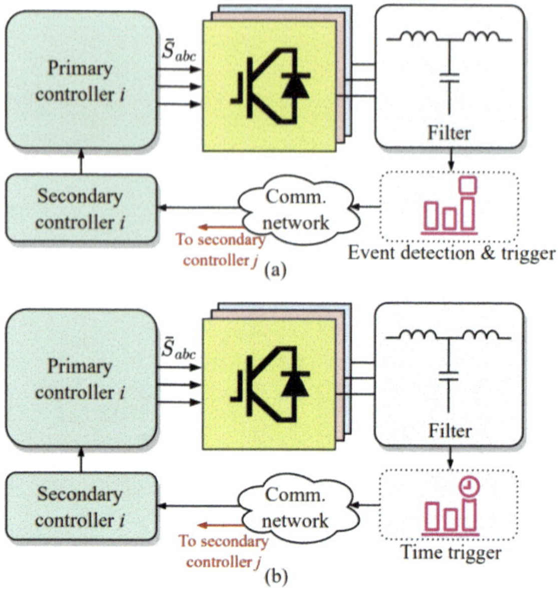

Fig. 4.7 Distributed control techniques. **a** Event-triggered distributed control. **b** Time-triggered distributed control

Table 4.1 Features of secondary control architectures

Architecture	Features	(Dis)Advantages
Centralized	• Central controller coordinates secondary frequency and voltage control • High fidelity communication channel required • Functions include: harmonic cancelation, unbalanced current reduction, and power quality management	• Performance and stability vulnerable to failure of single central controller • High cost due to communication infrastructure
Decentralized	• Each converter control module can implement secondary frequency and voltage control locally	• Low cost due to lack of communication technologies • Limited scale of application
Distributed	• Neighbor converter modules communicate over sparse network • Secondary frequency and voltage control implemented with limited data requirements	• Facilitates privacy of information • Robust to communication latency • Facilitates plug and play features • Can be scaled for wide-area systems

market forces and weather conditions. Therefore, this level has the slowest response time of several hours. Furthermore, the tertiary controller is also concerned with inter-microgrid interactions in microgrid clusters as well as their unified interaction with the main electric power grid.

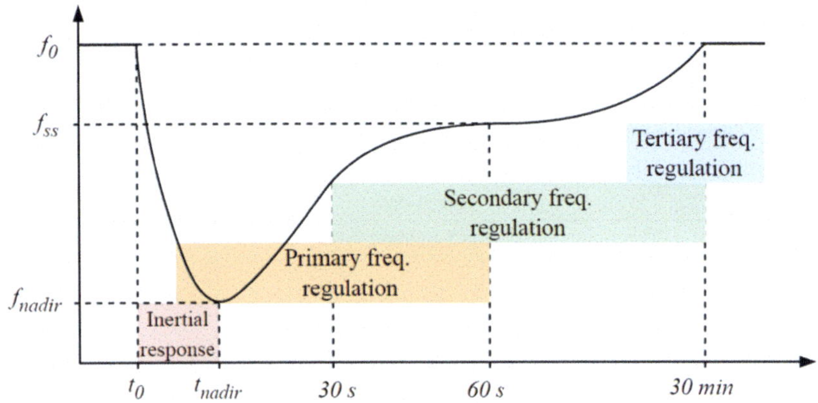

Fig. 4.8 Frequency regulation stages of the power system

Inertia Enhancement in Low-Inertia Power Systems

The power system frequency response is conventionally done in four stages: inertial response, primary regulation, secondary regulation, and tertiary regulation, as shown in Fig. 4.8. When a frequency disturbance event occurs at time t_0, inertial response is the first reaction to mitigate the disturbance. This is achieved by releasing the kinetic energy stored in the rotor mass for about 5 s. The inertial energy in the system determines how low the frequency nadir (f_{nadir}) can reach after the disturbance (higher f_{nadir} is desirable). Subsequently, the primary control by the governor is automatically triggered for about 30 s, when the frequency is expected to have been restored to a quasi-steady-state level (f_{ss}). Secondary control is also automatically triggered if the primary control does not correct the underfrequency situation. Finally, if the underfrequency persists, tertiary control is activated by manual control center intervention through the dispatch of power plants.

As earlier explained, converter-interfaced renewable sources lack the frequency regulation system of the conventional power system. Therefore, synthetic inertia in low-inertia power systems like microgrids has become essential in order to ensure the stability and safety of the system. In particular, the introduction of power-electronics-based inertia ensures that the ROCOF is limited, leading to better frequency containment within safe and regulatory limits. In recent times, power electronic converters are playing a helpful role by employing energy from wind turbines [18], VSGs [19], DC-link capacitors [20], ultracapacitors [21], and batteries [22], as shown in Table 4.2. Also, the cost of implementation varies with the source of energy that provides the virtual inertia, with wind turbine techniques and DC-link capacitors being the lowest cost options for practical implementation.

The swing equation provides a mathematical model of the operation of a synchronous generator connected to an electrical power network that has connected loads. The power difference between the generated and supplied load is equal to the rate of change of the kinetic energy stored in the rotating masses [23]:

Table 4.2 Techniques for inertia enhancement in power electronics systems

Methodology	Energy source	Cost of implementation
Wind turbines [18]	Mechanical	Low
Virtual synchronous machine [19]	Variable	Low to high
DC-link capacitors [20]	Electrical	Low
Ultracapacitors [21]	Electrical	High
Batteries [22]	Chemical	Highest

$$P_{\text{gen}} - P_{\text{load}} = \frac{dE_k}{dt} = \frac{d\left(\frac{1}{2}J\omega_g^2\right)}{dt} = J\omega_g\frac{d\omega_g}{dt} \tag{4.1a}$$

$$\frac{2H}{f}\frac{df}{dt} = \frac{P_{\text{gen}} - P_{\text{load}}}{S_g}, \tag{4.1b}$$

where P_{gen} is the power output of the synchronous generators, P_{load} is the power delivered to the load, E_k kinetic energy of the rotating masses, J is the total power system inertia, ω_g represents the power system frequency, and $\frac{df}{dt}$ represents the rate of change of frequency (ROCOF) of the power system. Equation (4.1b) is derived by re-writing (4.1a) in terms of frequency (Hz). It is desirable for the ROCOF of a power system to be very low. However, only electrical systems with very high inertia can resist sudden changes in frequency. Thus, in systems characterized by high penetration of converter-interfaced renewable energy sources, there is a need for the inertia of connected systems to be boosted by virtual inertia techniques.

4.3 Proposed Control Method

4.3.1 System Model

The LC filter is modeled as follows:

$$\frac{d}{dt}\begin{pmatrix} i_f \\ v_f \end{pmatrix} = \begin{bmatrix} -\frac{R_f}{L_f} & -\frac{1}{L_f} \\ \frac{1}{C_f} & 0 \end{bmatrix}\begin{pmatrix} i_f \\ v_f \end{pmatrix} + \begin{bmatrix} \frac{1}{L_f} & 0 \\ 0 & -\frac{1}{C_f} \end{bmatrix}\begin{pmatrix} v_i \\ i_o \end{pmatrix}, \tag{4.2}$$

where R_f, L_f, C_f, i_f, v_f represent the resistance, inductance, capacitance, current, and voltage of the filter, respectively.

Furthermore, the microgrid ensures power sharing by the classical droop equations:

$$\omega_{\text{ref}} = \omega^* - k_p(P - P^*) \tag{4.3a}$$

$$V_{\text{ref}} = V^* - k_q(Q - Q^*), \tag{4.3b}$$

where k_p and k_q are the $\omega - P$ and $V - Q$ droop coefficients respectively; P* and Q* are the reference active and reactive powers, respectively.

It was earlier noted that the conventional droop controller does not support inertia sufficiently. Hence, inertia emulation by the simplified swing equation is realized as follows [24]:

$$J\omega^* \frac{d\omega}{dt} = P_m - P - D(\omega - \omega^*), \tag{4.4}$$

where J and D represent the inertia and damping coefficients.

Furthermore, the MPC-based primary control has the control objective of ensuring that the predicted output voltage tracks its reference through the cost function in the following.

$$\begin{pmatrix} i_f(k+1) \\ v_f(k+1) \end{pmatrix} = \boldsymbol{A}' \begin{pmatrix} i_f(k) \\ v_f(k) \end{pmatrix} + \boldsymbol{B}' \begin{pmatrix} v_i(k) \\ i_o(k) \end{pmatrix}, \tag{4.5}$$

where $\boldsymbol{A}' = \boldsymbol{e}^{AT_s}$, $\boldsymbol{B}' = \int_0^{T_s} \boldsymbol{B} d\tau$, $\boldsymbol{A} = \begin{bmatrix} -\frac{R_f}{L_f} & -\frac{1}{L_f} \\ \frac{1}{C_f} & 0 \end{bmatrix}$, and $\boldsymbol{B} = \begin{bmatrix} \frac{1}{L_f} & 0 \\ 0 & -\frac{1}{C_f} \end{bmatrix}$.

Thus, the cost function is

$$J = \|v_f(k+1) - v_f^*\| + \lambda_i \|i_f(k+1) - i_f^*\|, \tag{4.6}$$

where λ_i is the current term weighting factor, and the reference current $i_f^* = \left(C_f \omega_{\text{ref}} v_{f\beta}^* + i_{o\alpha}\right) - jC_f \omega_{\text{ref}} v_{f\alpha}^* + ji_{o\beta}$.

4.3.2 MPC-Based Primary Control

The linear control of LC-filtered VSC in Fig. 4.9a is a common solution at the primary control level of microgrids. The overall control scheme results in a voltage-controlled voltage source. The outer loop is a basic droop P – f, Q – V controller, which is also effective in supporting grid-forming control of the VSC. The inner loop consists of voltage and current controllers $G_v(s)$, $G_i(s)$ in cascade, respectively. $Gi(s)$ operates at a faster bandwidth than $G_v(s)$. The output of $G_i(s)$ drives the pulse width modulator (PWM) to regulate the average output voltage during each sample period.

The limitation of linear control methods is also shown by the bandwidth scale of the cascaded control blocks in Fig. 4.10. Assuming that the current loop operates at a decade lower than the sampling frequency f_s (converter switching frequency is f_{sw}), to facilitate decoupling of the control loops, the voltage loop will operate at a decade

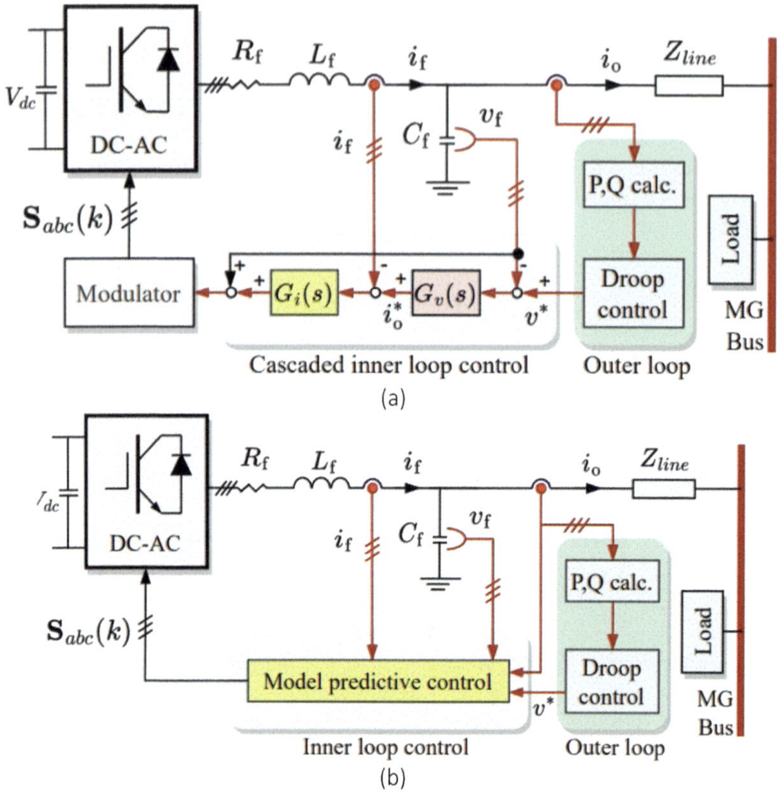

Fig. 4.9 Microgrid VSC and its control. **a** Conventional linear control. **b** FCS-MPC scheme

lower. Thus, the overall system's bandwidth will be about two orders of frequency lower than fs, resulting in a slower dynamic response.

Microgrids operate in multiple hierarchies, with the VSC control being the primary control layer. Therefore, if the primary control can be designed with faster bandwidth, the secondary and tertiary hierarchies will be capable of operating with improved dynamic response. MPC facilitates MIMO control of nonlinear systems and overcomes the limitations of cascaded linear control. This is beneficial for microgrids which require multi-hierarchical control loops having different time scales. MPC can help improve the overall dynamic response. Figure 4.9b shows that finite control-set MPC (FCS-MPC) offers faster dynamic response because it can function without a dedicated modulator stage. The bottom two bandwidth scale plots of Fig. 4.10 show that nonlinear control (in this case, MPC) that obviates a dedicated PWM stage can increase the VSC's primary control bandwidth by an order of up to two decades.

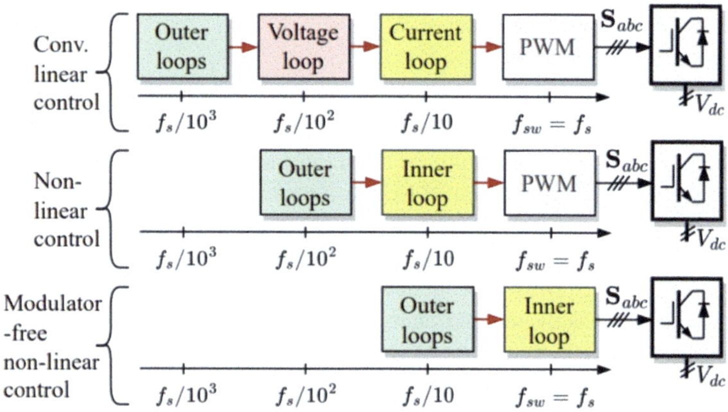

Fig. 4.10 Bandwidths of control loops for VSC

4.3.3 Distributed Consensus-Based Secondary Control

The secondary controller in this study is a proportional-integral controller defined as follows:

$$\delta_{\omega} = \left(K_{p\omega} + \frac{K_{i\omega}}{s}\right)(\omega^{*} - \overline{\omega}) \tag{4.7}$$

$$\delta_{v} = \left(K_{pV} + \frac{K_{iV}}{s}\right)\left(V^{*} - \overline{V}\right), \tag{4.8}$$

where δ_{ω}, and δ_{v} are the secondary compensatory signals, and $K_{py,}K_{iy}$ are the secondary PI-controller coefficients $\forall y \in \{\omega, V\}$. In order to ensure effective control with minimum communication resources (only neighbor converters need to share data with each other), while ensuring increased privacy for each agent, the average terms $\overline{\omega}$ and $\overline{V}$ are derived from an average consensus observer defined below [10].

$$\begin{cases} \dot{\Delta}_i = -\gamma\Delta_i - \sum\limits_{j\in N_i} f_{ij}\left(\overline{y}_i - \overline{y}_j\right) + \sum\limits_{j\in N_i} f'_{ij}\left(g_i - g_j\right) \\ \dot{g}_i = f'_{ij}\left(\overline{y}_i - \overline{y}_j\right), \end{cases} \tag{4.9}$$

where $\Delta_{\mathrm{i}} = y_i - \overline{y}_i$, $\gamma \geq 0$ is a global observer for $y \in \{\omega, V\}$, g is a stability term for reinitialization, and f'_{ij} represents terms of the adjacency matrix.

4.4 Validation and Discussion

The overall proposed distributed secondary control scheme in Fig. 4.11 was implemented on the test microgrid shown in Fig. 4.12 with parameters shown in Table 4.3. The dynamic observer in (4.9) was used to estimate the dynamic average parameters. Simulations were carried out with PLECS software to validate the control scheme.

Primary Frequency Response

Figure 4.13 shows the primary frequency with two control methods. A 70% increase of active power was activated and later deactivated after 0.06 s. The conventional droop produced a high change of frequency (0.19 Hz) due to low inertia in the microgrid. However, when virtual inertia was introduced from the VSG, the resulting reduced ROCOF produced a lower frequency change of 0.02 Hz. This represents a reduction of 89.5% in the change of frequency. The impact of VSG control in limiting the ROCOF is shown in Fig. 4.14 for the case of one VSC. The maximum ROCOF during transient response is decreased from a maximum of 4.1 Hz/s (with

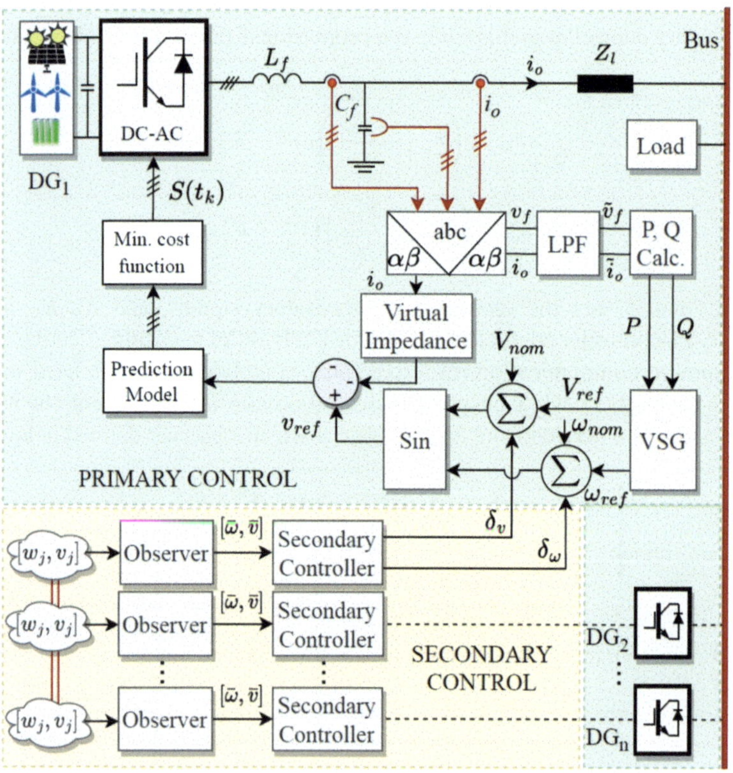

Fig. 4.11 Microgrid system with distributed DERs and proposed control system

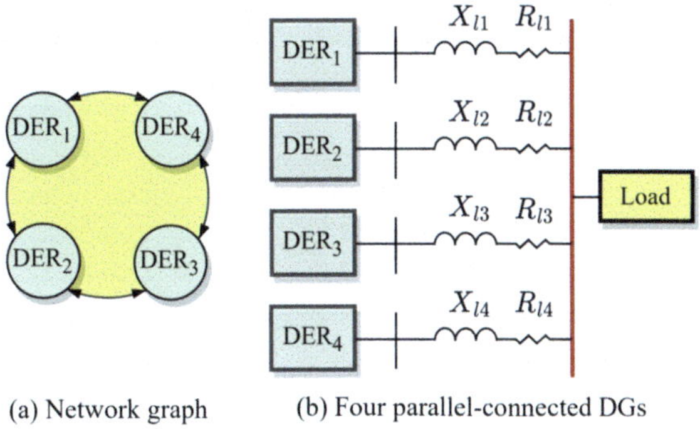

Fig. 4.12 Test microgrid system

Table 4.3 Key electrical and control parameters of the microgrid

Parameter	Symbol	Value
DC-link voltage	V_{dc}	500 V
Nominal frequency	f^*	50 Hz
Nominal voltage	V^*	250 V
Filter parameters	R_f, L_f, C_f	$R_f = 0.1\Omega$, $L_f = 2.4$ mH, $C_f = 25\ \mu$F
Line impedance	R_l, L_l	$R_l = 0.73\Omega$, $L_l = 13.33$ mH
Droop coefficients	k_p, k_q	DER1, DER3: 2.1 mV/Var, 5.2 mrad/W; DER2, DER4: 3.15 mV/Var, 5.2 mrad/W
Moment of inertia	J	DER1, DER3: 0.32 kg m^2; DER2, DER4: 0.64 kg m^2
Weighting factor	λ_i	5.0
Secondary controller	$K_{p\omega}, K_{i\omega}, K_{pV}, K_{iV}$	7.5×10^3, 0.15, 4.0×10^3, 10.0×10^3
Communication rate	T_s	10 μs

conventional droop control) to 1.0 Hz/s (for VSG control), showing the effectiveness of decreasing ROCOF at the primary level.

Secondary Frequency and Voltage Control

Despite the improvement in primary frequency response by the introduction of virtual inertia, the frequency response can be further improved by distributed consensus secondary control earlier discussed in this chapter. Figures 4.15 and 4.16 show the results of the proposed secondary control on the test system. First, Fig. 4.15 shows that the secondary control compensated for the frequency deviation at the primary level by maintaining the frequency at the nominal value even during the change in active power. Similarly, the primary voltage deviation of 2.4 V arising from reactive power change is effectively compensated by the secondary voltage controller (Fig. 4.16).

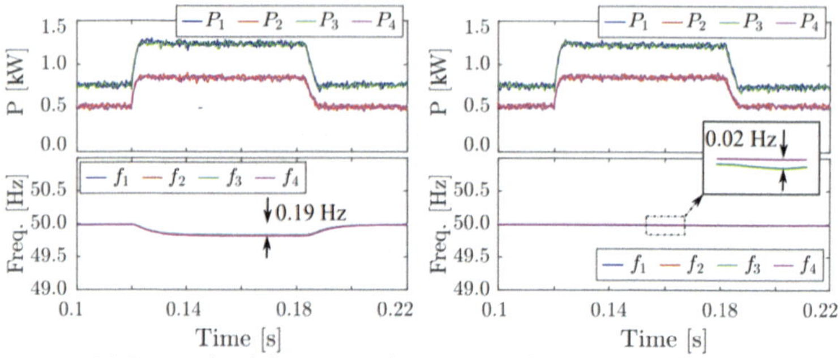

(a) Conventional droop control (b) Proposed primary control

Fig. 4.13 Primary frequency control performance

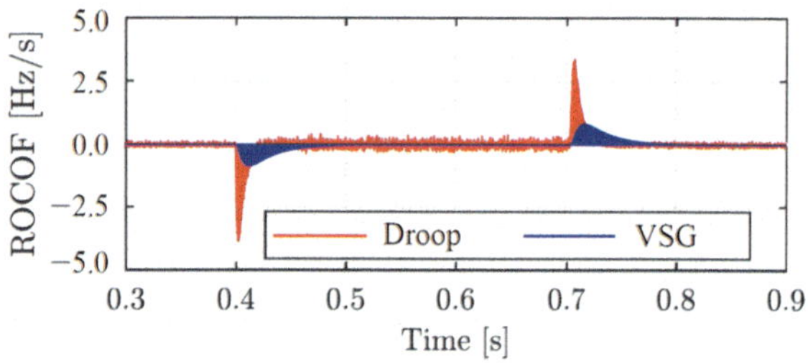

Fig. 4.14 ROCOF of primary frequency control

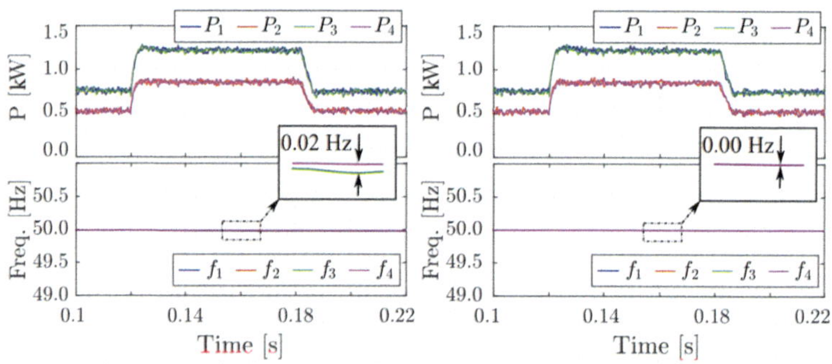

(a) Proposed primary control (b) Proposed secondary control

Fig. 4.15 Secondary control of frequency

Communication Delay

The impact of communication delay was tested by comparing the nominal communication update rate of 10 μs with slower rates of 100 μs and 1 ms. The results in Fig. 4.17 show that despite decreasing the communication update rate by two orders of magnitude, the proposed method is still effective in active power sharing.

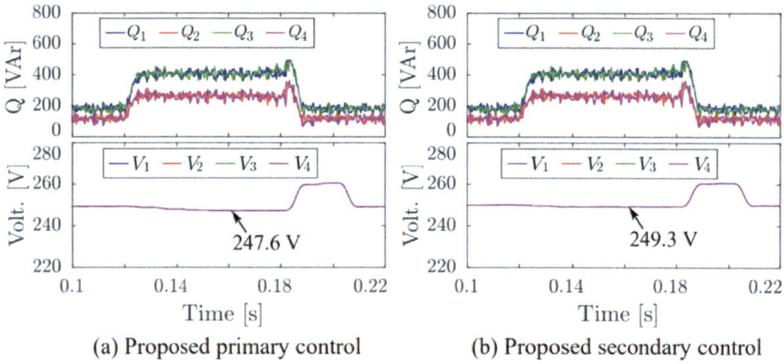

Fig. 4.16 Secondary control of voltage

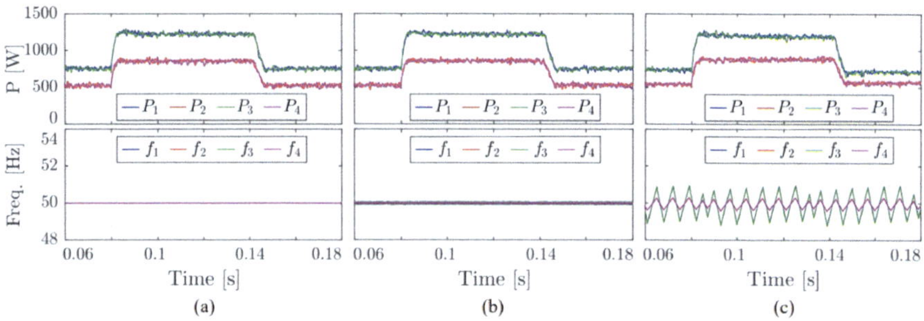

Fig. 4.17 Test of communication delay. **a** 10 μs, **b** 100 μs, and **c** 1.0 ms refresh rate

Communication Failure

The case of communication failure between DER1 and DER2 is shown in Fig. 4.18. The ring communication topology still provides a means for the data to/from DER2 to be communicated through the healthy part of the network. Thus, the control output was unaffected by the break in the network.

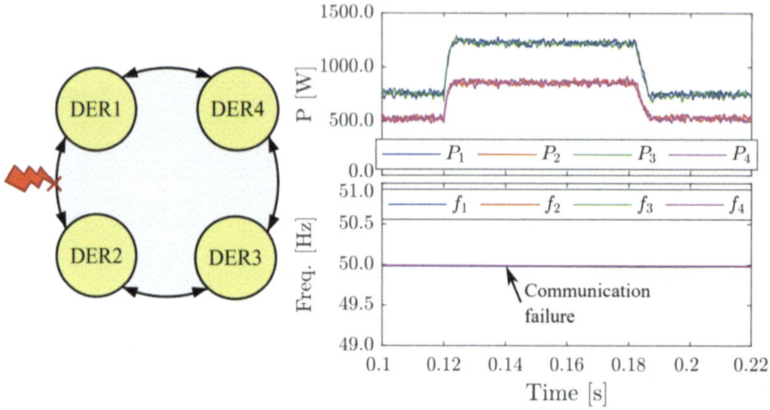

Fig. 4.18 Test of communication failure

4.5 Conclusion

In this chapter, the problem of low inertia of converter-interfaced renewable sources in microgrids was studied. Existing techniques to overcome this challenge through hierarchical control were surveyed. Finally, a new solution was demonstrated by complementing the inertia emulation of VSG with the fast-dynamic properties of MPC in the distributed secondary regulation of frequency and voltage for an AC microgrid. The proposed distributed control scheme accurately regulates both the secondary frequency and voltage deviations in the primary control level. The operation of the consensus-based average observer guarantees robust control performance under conditions of communication latency and failure.

References

1. W. Hu, Y. Lv, Y. Shen, X. Dou, F. Yang, Inertia-enhanced method for active distribution network based on hierarchical control structure. Int. J. Electr. Power Energy Syst. **142**, 108365 (2022)
2. J. Fang, H. Deng, N. Tashakor, F. Blaabjerg, S.M. Goetz, State-space modeling and control of grid-tied power converters with capacitive/battery energy storage and grid-supportive services. IEEE J. Emerg. Sel. Top. Power Electron. (2021)
3. O. Stanojev, U. Markovic, P. Aristidou, G. Hug, D. Callaway, E. Vrettos, MPC-based fast frequency control of voltage source converters in low-inertia power systems. IEEE Trans. Power Syst. **37**(4), 3209–3220 (2020)
4. K. Kumar, S. Bae, Coordinated dynamic power management for renewable energy-based grid-connected microgrids using model predictive control. IEEE Trans. Industr. Inf. **19**(9), 9596–9608 (2023). https://doi.org/10.1109/TII.2022.3231409
5. U. Tamrakar, T.M. Hansen, R. Tonkoski, D.A. Copp, Model predictive frequency control of low inertia microgrids, in *2019 IEEE 28th International Symposium on Industrial Electronics (ISIE)* (IEEE, June 2019), pp. 2111–2116

6. R. Heydari, Y. Khayat, M. Naderi, A. Anvari-Moghaddam, T. Dragicevic, F. Blaabjerg, A decentralized adaptive control method for frequency regulation and power sharing in autonomous microgrids, in *2019 IEEE 28th International Symposium on Industrial Electronics (ISIE)* (IEEE, June 2019), pp. 2427–2432
7. O. Babayomi, Z. Zhang, Y. Li, R. Kennel, Adaptive predictive control with neuro-fuzzy parameter estimation for microgrid grid-forming converters. Sustainability **13**(13), 7038 (2021)
8. O. Babayomi, Y. Li, Z. Zhang, Distributed consensus-based reactive power sharing in microgrids: a predictive virtual capacitance control technique. Int. J. Electr. Power Energy Syst. **141**, 108139 (2022)
9. O. Babayomi, Y. Li, Z. Zhang, R. Kennel, J. Kang, Overview of model predictive control of converters for islanded AC microgrids, in *2020 IEEE 9th International Power Electronics and Motion Control Conference (IPEMC2020-ECCE Asia).* (IEEE, November 2020), pp. 1023–1028
10. O. Babayomi, Z. Li, Z. Zhang, Distributed secondary frequency and voltage control of parallel-connected vscs in microgrids: a predictive VSG-based solution. CPSS Trans. Power Electron. Appl. **5**(4), 342–351 (2020)
11. J.F. Chen, C.-L. Chu, Combination voltage-controlled and current-controlled PWM inverters for UPS parallel operation. IEEE Trans. Power Electron. **10**(5), 547–558 (1995)
12. Z. Zhang, O. Babayomi, T. Dragicevic, R. Heydari, C. Garcia, J. Ro-driguez, R. Kennel, Advances and opportunities in the modelpredictive control of microgrids: Part i–primary layer. Int. J. Electr. Power Energy Syst. **134**, 107411 (2022)
13. Z. Shi, J. Li, H.I. Nurdin, J.E. Fletcher, Comparison of virtual oscillator and droop controlled islanded three-phase microgrids. IEEE Trans. Energy Convers. **34**(4), 1769–1780 (2019)
14. M.A. Awal, H. Yu, H. Tu, S.M. Lukic, I. Husain, Hierarchical control for virtual oscillator-based grid-connected and islanded microgrids. IEEE Trans. Power Electron. **35**(1), 988–1001 (2020)
15. S. Luo, W. Wu, E. Koutroulis, H.S.-H. Chung, F. Blaabjerg, A new virtual oscillator control without third-harmonics injection for DC/AC inverter. IEEE Trans. Power Electron. **36**(9), 10879–10888 (2021)
16. J. Hu, Y. Shan, K.W. Cheng, S. Islam, Overview of power converter control in microgrids—challenges, advances, and future trends. IEEE Trans. Power Electron. **37**(8), 9907–9922 (2022). https://doi.org/10.1109/TPEL.2022.3159828
17. O. Babayomi, Z. Zhang, T. Dragicevic, R. Heydari, Y. Li, C. Garcia, J. Rodriguez, R. Kennel, Advances and opportunities in the modelpredictive control of microgrids: Part ii–secondary and tertiary layers. Int. J. Electr. Power Energy Syst. **134**, 107339 (2022)
18. J. Ma, Y. Gu, Y. Shen, Y. Zhou, P. Cheng, Stability analysis of power grid connected with direct-drive wind farm containing virtual inertia based on integrated dissipation energy model. IEEE Trans. Sustain. Energy **12**(4), 2378–2392 (2021). https://doi.org/10.1109/TSTE.2021.3095171
19. J. Fang, R. Zhang, H. Li, Y. Tang, Frequency derivative-based inertia enhancement by grid-connected power converters with a frequency-locked-loop. IEEE Trans. Smart Grid **10**(5), 4918–4927 (2019)
20. Q. Peng, J. Fang, Y. Yang, T. Liu, F. Blaabjerg, Maximum virtual inertia from DC-link capacitors considering system stability at voltage control timescale. IEEE J. Emerg. Sel. Top. Circuits Syst. **11**(1), 79–89 (2021). https://doi.org/10.1109/JETCAS.2021.3049686
21. M. Hajiakbari Fini, M.E. Hamedani Golshan, J.R. Martí, A. Ketabi, Determining the required frequency control reserve and capacity and location of synchronous and virtual inertial resources. IEEE Trans. Sustain. Energy **14**(1), 27–38 (2023). https://doi.org/10.1109/TSTE.2022.3198695
22. A. Jain, N.P. Padhy, M.K. Pathak, Cost-effective size of HESS incorporated in VSG with enhanced performance of battery providing implicit inertia. IEEE Trans. Ind. Appl. https://doi.org/10.1109/TIA.2023.3275930

23. K. Yan, G. Li, R. Zhang, Y. Xu, T. Jiang, X. Li, Frequency control and optimal operation of low-inertia power systems with HVDC and renewable energy: a review. IEEE Trans. Power Syst. https://doi.org/10.1109/TPWRS.2023.3288086
24. Z. Zhang, et al., An adaptive virtual synchronous generator based model predictive control with enhanced frequency support capability in micro-energy systems, in *2023 IEEE International Conference on Predictive Control of Electrical Drives and Power Electronics (PRECEDE), Wuhan, China* (2023), pp. 1–6. https://doi.org/10.1109/PRECEDE57319.2023.10174514

Chapter 5
Power Quality and Stability—Control Strategy and Grid Impedance Detection of Three-Phase Grid-Connected Inverters Under Weak Power Grids

Yong Yang

Abstract In recent years, with the rapid development of power electronics technologies, a significant number of power electronic devices are connected to low voltage distribution power networks. As a result, a large number of nonlinear loads are introduced into the power grid, which brings harmonic pollution to the power grid and makes the power grid environment worse. With the rapid application of distributed generation systems in the power grid, the research of grid-connected inverters has attracted increasing attention, and researchers worldwide have conducted in-depth research on the modeling, control strategy, and other aspects of grid-connected inverters. This chapter analyzes the control structure and mathematical model of LCL three-phase grid-connected inverter with inverter current feedback control. Then, the capacitor voltage Lead compensation feedforward method and the point of common coupling (PCC) voltage control reactive power compensation method are used to solve the LCL resonance problem and the PCC voltage drop problem when the three-phase grid-connected inverter operates under a weak power grid. The effectiveness of the proposed scheme is verified through comparative simulation experiments in MATLAB/Simulink, achieving the goal of stable operation of PV grid-connected inverters under weak power grid conditions of short circuit ratio (SCR). Furthermore, this chapter proposes a new grid impedance detection method incorporating the complex coefficient filter (CCF) with full-order capacitor current observer for a T-type three-level grid-connected inverter controlled by the inverter output current feedback. Compared with conventional CCF impedance detection algorithms, the proposed method reduces the number of current sensors and detects the grid impedance accurately. First, based on the sampled inverter output current and grid-connected voltage signals, the grid-connected current is estimated. Then, the CCF method is used to extract harmonics from the grid-connected current and

Y. Yang (✉)
School of Rail Transportation, Soochow University, Suzhou 215131, P. R. China
e-mail: yangy1981@suda.edu.cn

J. Hu (ed.), *Community Energy and Microgrids*, Green Energy and Technology,
https://doi.org/10.1007/978-981-97-6297-2_5

voltage signals to calculate the grid impedance. Finally, the correctness of the full-order capacitor current observer is verified by simulation and the feasibility and effectiveness of the proposed algorithm are verified experimentally based on a laboratory prototype.

5.1 Research Status of Control Methods of Grid-Connected Inverters Under Weak Power Grid

In recent years, with the rapid development of power electronics technologies, a significant number of power electronic devices are connected to low voltage distribution power networks. As a result, a large number of nonlinear loads are introduced into the power grid, which brings harmonic pollution to the power grid and makes the power grid environment worse. In countries such as China and Australia, clean energy sources such as solar and wind power that can be developed and utilized are mainly distributed in remote areas and regional communities. Therefore, there will be a large distance between the distributed generation system in remote areas and the peak electricity consumption area. In order to transmit the produced electrical energy, it is necessary to pass through long transmission lines, and a large number of transformers will be introduced, resulting in the introduction of a certain line equivalent impedance between the grid-connected inverter and the power grid. For instance, a new 190 km overhead electricity transmission line is being proposed to unlock the renewable energy potential of western Victoria as a key Renewable Energy Zone and will help to deliver clean and affordable energy to Melbourne. The main components of the equivalent impedance of a transmission line are the resistor and the inductor. In this case, the power grid cannot be equivalent to the ideal voltage source, and it is called a weak power grid. Its characteristics are as follows: there is an impedance in the power grid that cannot be ignored, and the impedance value will fluctuate. In the case of a weak power grid, the operating state of the grid-connected inverter is not the same as ideal power grid conditions due to the non-negligible grid impedance. Understanding the influence of grid impedance on system performance such as dynamic performance and stability of grid-connected inverters is crucial to ensure their normal and stable operation under weak power grid conditions [1–5]. With the rapid application of distributed generation systems in the power grid, the research of grid-connected inverters has attracted increasing attention, and researchers worldwide have conducted in-depth research on the modeling, control strategy, and other aspects of grid-connected inverters. However, most of the studies are based on ideal power grid conditions, and there are few studies on weak power grid conditions in community microgrids. Since the characteristics of a weak power grid are becoming more and more obvious in the actual power grid, it is very necessary to study the control strategy of grid-connected inverters under a weak power grid. In December 2019, the Chinese standard GB 38,755–2019 "Guidelines for Safety and Stability of Power Systems" was released, in which the "Short Circuit

Ratio" (SCR) was used for the first time to characterize the power grid strength of the grid-connected points of new energy power stations:

$$\mathrm{SCR} = \frac{S_{ac}}{P_n} = \frac{U_n^2/\sum Z}{P_n}, \quad (5.1)$$

where S_{ac} is the short-circuit capacity of the power grid, P_n is the rated active power, Z is the equivalent impedance of the AC system, and U_n is the rated voltage of the power grid. The smaller the SCR is, the weaker the power grid is. It is generally considered the power grid is a weak power grid when $2 < \mathrm{SCR} < 3$, and the power grid is an extremely weak power grid when $\mathrm{SCR} < 2$. The certification standard for stable operation of grid-connected inverters under weak power grid conditions has been set by the China Electric Power Research Institute as SCR = 1.5. Numerous new energy enterprises in China have conducted research on this issue. The 2019 China Huawei report shows that Huawei's intelligent photovoltaic (PV) inverters have realized the stable operation of SCR lower limit equal to 1.5, which is in an absolute leading position in the industry. In 2020, the entire series of products of China SINENG Electric Appliances passed the weak power grid-connection performance test of the Chinese Academy of Electrical Sciences, achieving the standard of stable machine operation under SCR = 1.5. In 2021, a report issued by China SUNGROW Company showed that the products, centralized inverters, and string inverters, have passed certification tests with SCR = 1.02 and SCR = 1.16, respectively, setting a new record for weak power grid adaptability of inverters in the industry.

The control methods of grid-connected inverters are mainly divided into two modes: current control and voltage control [6]. The current-type inverters generally adopt a double closed-loop control method, which calculates the given current amplitude through the DC voltage control and power control of the outer loop, calculates the given current angle through the phase-locked loop (PLL) tracking the frequency and phase of the point of common coupling (PCC) voltage, calculates the modulation voltage through the current inner loop, and then processes the modulation module to obtain the control signal of the switching device. The voltage-type inverters generally do not need PLL. They use power control to calculate the given voltage amplitude and system phase in the inner loop. Traditional power control methods include droop control and virtual synchronous generator (VSG) [7], and then control power electronic switching devices through the inner loop and modulation module. The specific control block diagrams are shown in Figs. 5.1 and 5.2.

In addition to structural differences, the grid-connected inverters with two control modes also have the following differences:

(1) The DC side of the current-type inverter can be regarded as a constant current source, the output current of the inverter is a rectangular wave, which is independent of the load impedance angle, and the output voltage is approximately seen as a sine wave. The DC side of the voltage-type inverter can be regarded as a constant voltage source, the output voltage of the inverter is a rectangular

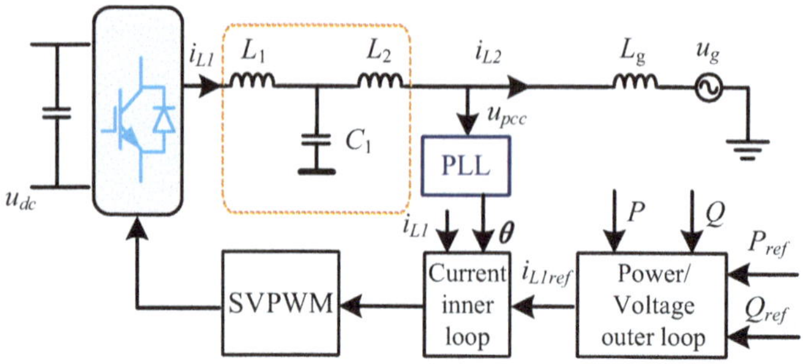

Fig. 5.1 Control block diagram of the current-type grid-connected inverter

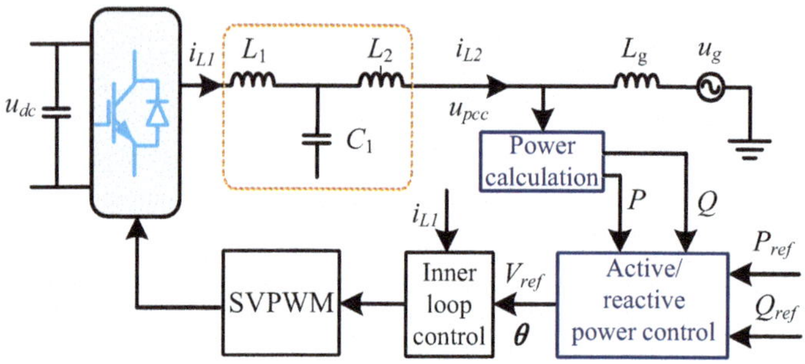

Fig. 5.2 Control block diagram of the voltage-type grid-connected inverter

wave, which is independent of the load impedance angle, and the output current is approximately seen as a sine wave.

(2) The current-type inverters operate in grid-connected mode, while the voltage-type inverters can operate both in grid-connected conditions and island mode.

(3) The current-type inverter is more stable under strong power grid conditions, and the voltage-type inverter is more adaptable under weak power grid conditions.

In addition to the above current-mode and voltage-mode control methods, there are other control strategies, such as model predictive control (MPC), repetitive control, sliding mode control (SMC), robust control, and neural network control [8–10]. These control methods have their own advantages and disadvantages. For example, MPC has good dynamic performance of the inverter, but the calculation is large and the inverter performance is affected by inverter parameters. At present, the most commonly used control methods are current-mode control and voltage-mode control, and the current-mode control method is more widely used.

In order to solve the LCL resonance problem of the grid-connected inverters operating under weak power grids, commonly resonance suppression methods are divided into passive damping and active damping [11–13]. The basic form of passive damping is to connect resistors in series or parallel with the LCL filter. From the analysis of frequency characteristics, it can be seen that the damping effect of the resistor in parallel with the LCL filter is better. In recent years, domestic and foreign scholars have proposed improved passive damping methods, such as adding bypass inductor, adding bypass inductor and capacitor, splitting capacitor, splitting capacitor, and adding bypass inductor. These methods can reduce the loss of damping resistance and have strong attenuation ability of high-frequency components of harmonics, but will make the circuit structure complex and increase the cost and volume of the system [11–13].

In order to avoid the negative impact of passive damping methods, domestic and foreign scholars have proposed active damping methods, which add resonance peak suppression algorithms to the control loop to achieve resonance suppression. The active damping methods currently studied mainly include capacitor current proportional feedback, capacitor voltage first differential feedback, and grid current second differential feedback. The principle is to simulate passive damping through state variable feedback control, which is equivalent to the effect of "virtual impedance". Generally speaking, differential processing of the signals amplifies high-frequency noise and weakens the damping effect. However, the capacitor current is a high-frequency signal, which is difficult to accurately estimate and there will be errors in sampling values [11–13]. Reference [14] proposed an active damping method based on a notch filter, but this kind of method needs to know the specific parameters, and the practical application is more complex. Compared with feedback control, reasonable feedforward control can be unaffected by system lag and will improve the response speed of the system. In addition, when the power grid is very weak, the ordinary feedforward cannot fully compensate for the system phase. The approximate differential process of the Lead compensation link and the Lead-lag compensation link are introduced, and good results are achieved [15]. By comparison, the parameter determination of Lead compensation is simpler. In summary, this chapter adopts the active damping method of capacitor voltage Lead compensation feedforward to suppress LCL resonance.

To obtain higher electric energy transmission efficiency and economic benefits, the grid-connected inverters usually operate at unit power factor (PF). However, according to the theoretical analysis in literature, the minimum Short Circuit Ratio (SCR) accepted by the inverter under unit power factor is 2. Therefore, if the grid-connected inverter operates under extremely weak power grid conditions, reactive power compensation needs to be carried out on the system, otherwise it will cause voltage fluctuations, reduce power quality, and even lead to the system off-grid. Therefore, it is necessary to study the reactive power compensation methods of grid-connected inverters.

At present, power electronic reactive power compensation equipment based on high voltage direct current (HVDC) and flexible alternative current transmission systems (FACTS) has been used to improve the electric energy transmission capacity of weak power grid. However, the installation of these power electronic devices poses higher demands on system stability and power quality. Therefore, it is becoming increasingly attractive to utilize the power controllability of grid-connected inverters to improve the electric energy transmission capacity under weak power grid conditions, which is also beneficial to reduce the cost of grid infrastructure upgrades.

The photovoltaic power generation system specified in China should operate within the PF range of ±0.9/0.95. Many literatures have proposed reactive voltage control strategies of grid-connected inverters, such as reactive voltage control based on PCC voltage, constant reactive power control, constant power factor control, and so on. Because it is difficult to accurately predict the photovoltaic (PV) curve and load demand curve to calculate the amount of reactive power compensation, and the reactive power of the inverter is proportional to the active power, which will produce excess reactive power and increase the loss, this chapter uses the control method to determine the amount of reactive power compensation based on the offset of the PCC voltage.

5.2 Control Structure and Mathematical Model of Three-Phase Grid-Connected Inverters

5.2.1 Control Structure of Three-Phase Grid-Connected Inverters

Figure 5.3 shows the control block diagram of LCL three-phase grid-connected inverters with inverter current feedback control, where u_{dc} is the DC power supply, C_{n1} and C_{n2} are the dc-link upper and lower capacitors, respectively, L_1, C_1 and L_2 are the LCL filters, L_g represents the grid impedance, i_{L1} is the inverter output current, u_i is the inverter output voltage, i_{L2} is the grid-connected current, u_{pcc} is the point of common coupling (PCC) voltage, u_g is the grid voltage, u_c and i_c are the capacitor voltage and current, respectively. Its working principle is to obtain the reference current i_{ref} based on the phase-locked angle calculated from the PCC voltage and the amplitude calculated from the voltage outer loop, then compare it with the sampled inverter current i_{L1} to obtain the error signal, and then pass through the proportional resonance (PR) controller, in order to suppress LCL resonance, capacitor voltage Lead compensation feedforward signal is also added to obtain the modulation signal. Finally, the control signals of the switching devices are obtained through space vector pulse width modulation (SVPWM).

According to Kirchhoff's law, the state equations of the system shown in Fig. 5.3 can be obtained:

$$\begin{cases} L_1 \dfrac{di_{L_1abc}}{dt} = u_{iabc} - u_{cabc} \\ C_1 \dfrac{du_{cabc}}{dt} = i_{L_1abc} - i_{L_2abc} \\ L_2 \dfrac{di_{L_2abc}}{dt} = u_{cabc} - u_{pccabc} \end{cases} . \tag{5.2}$$

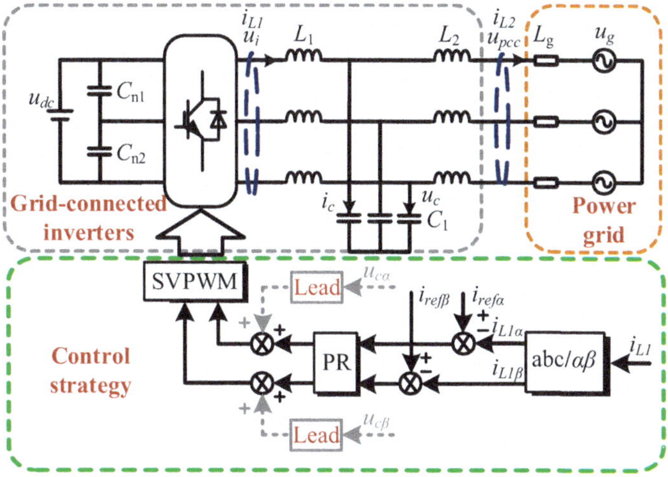

Fig. 5.3 Control block diagram of LCL three-phase grid-connected inverter with inverter current feedback control

In order to simplify the computational complexity, Clarke transformation is used to transform the variables from the three-phase stationary coordinate system to the two-phase stationary coordinate system. The specific transformation formula is

$$\begin{bmatrix} x_\alpha \\ x_\beta \end{bmatrix} = m \begin{bmatrix} 1 & -\frac{1}{2} & -\frac{1}{2} \\ 0 & \frac{\sqrt{3}}{2} & -\frac{\sqrt{3}}{2} \end{bmatrix} \begin{bmatrix} x_a \\ x_b \\ x_c \end{bmatrix}. \tag{5.3}$$

The coefficient m is selected based on the transformation conditions, when $m = 2/3$ is equal amplitude transformation, when $m = \sqrt{2/3}$ is equal power transformation.

To analyze the system in the frequency domain, we need to perform Laplace transform on formula (5.2):

$$\begin{cases} sL_1 i_{L_1}(s) = u_i(s) - u_c(s) \\ sC_1 u_c(s) = i_{L_1}(s) - i_{L_2}(s) \\ sL_2 i_{L_2}(s) = u_c(s) - u_{pcc}(s) \end{cases}. \tag{5.4}$$

5.2.2 Mathematical Model of Three-Phase Grid-Connected Inverter

According to the control block diagram in Fig. 5.3, the LCL three-phase grid-connected inverter is modeled mathematically. The specific model is shown in Fig. 5.4.

G_L (s) in Fig. 5.4 is a low-pass filter (LPF) with a cutoff frequency of 4 kHz, which is used to filter out high-frequency switching harmonics:

$$G_L = \frac{8000\pi}{s + 8000\pi}. \tag{5.5}$$

G_c (s) is current inner loop PR controller:

$$G_c(s) = k_p + \frac{2k_r\omega_i s}{s^2 + 2\omega_i s + \omega_0^2}, \tag{5.6}$$

where k_p and k_r are PR controller coefficients, ω_i is the cutoff frequency of PR controller, $\omega_0 = 314.16$ rad/s. k_{pwm} is the modulation coefficient, $G_d = e^{-1.5T_s s} \approx \frac{1-1.5s/32000}{1+1.5s/32000}$ is the control delay, H_i and H_v are the current and voltage sampling coefficient, T_2 is the capacitor voltage Lead compensation link. u_{pcc} is regarded as the disturbance term, namely $u_{pcc} = 0$. According to the Fig. 5.4, the open-loop transfer function of the current inner loop is expressed as

$$G(s) = \frac{i_{L1}(s)}{i_{ref}(s)} = \frac{G_L G_c k_{pwm} G_d (1 + s^2 C_1 L_2)}{s^3 L_1 C_1 L_2 - s L_2 T_2 H_v k_{pwm} G_d + s(L_1 + L_2)}. \tag{5.7}$$

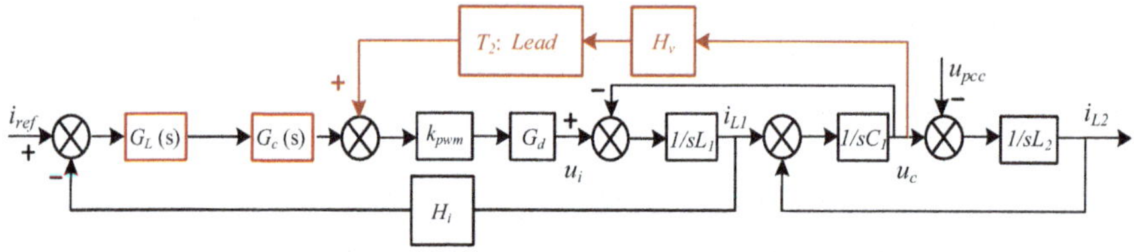

Fig. 5.4 Mathematical model of LCL three-phase grid-connected inverters

5.2.3 *Capacitor Voltage Lead Compensation Feedforward Strategy*

LCL resonance problem of three-phase grid-connected inverters under weak power grid

When the capacitor voltage Lead compensation feedforward strategy under weak power grid is not considered, which is $T_2 = 0$, the current inner loop open-loop bode diagram drawn by formula (5.7) through MATLAB is shown in Fig. 5.5, and the simulation parameters of the three-phase grid-connected inverters are shown in Table 5.1.

In Fig. 5.5, the grid impedance $L_g = 10\text{mH}$, SCR = 2.312 at rated power 20 kW, and it is in a weak power grid. According to the bode diagram, the gain margin (GM) is equal to −52 dB, the phase margin (PM) is equal to 8.91 deg, and the GM is less than 0. At this point, the system is in an unstable state, and the unstable frequency is

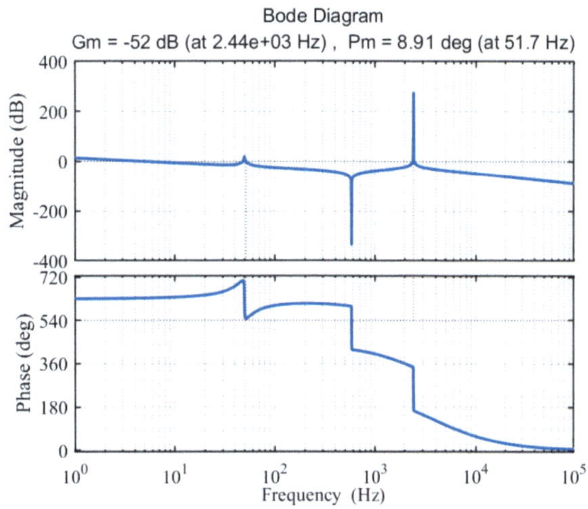

Fig. 5.5 Current inner loop open-loop bode diagram without capacitor voltage lead compensation feedforward strategy

Table 5.1 Simulation parameters of the three-phase grid-connected inverters

Parameter	Value	Parameter	Value
DC Voltage u_{dc}	800 V	Bus capacitor C_{n1}, C_{n2}	150 μF
Power grid voltage u_g	220 V	PR controller: k_p	6
Fundamental frequency f	50 Hz	PR controller: k_r	600
Sampling frequency f_s	16 kHz	PR controller cutoff frequency: ω_i	1
Filter inductor L_1	0.6 mH	Voltage sampling coefficient H_v	16
Filter capacitor C_1	7.5 μF	Current sampling coefficient H_i	16
Filter inductor L_2	0.05 mH	Rated power	20 kW

2.44 kHz, which is equal to the resonant frequency $f_r = \frac{1}{2\pi}\sqrt{\frac{L_1+(L_2+L_g)}{L_1C_1(L_2+L_g)}} = 2440$ Hz. It can be seen from Fig. 5.5 that in order to ensure stable operation of the three-phase grid-connected inverter, it is necessary to solve the problem of LCL resonance at high impedance.

Figure 5.6 shows the simulation results of the inverter current i_{L1} and Fast Fourier Transform (FFT) analysis of the inverter current i_{L1} in MATLAB/Simulink with the simulation parameters listed in Table 5.1. From the Fast Fourier Transform (FFT) analysis of the inverter current i_{L1} shown in Fig. 5.6, it can be seen that the three-phase grid-connected inverter system is unstable, and the harmonics mainly concentrated at 2475 Hz, which is consistent with theoretical analysis.

Capacitor voltage Lead compensation method

To solve the problem of LCL resonance in a three-phase grid-connected inverter under a weak power grid, this chapter adopts the active damping method of capacitor voltage Lead compensation feedforward, which adds capacitor voltage through the Lead link to the modulation wave to suppress LCL resonance under weak power grid. The specific control block diagram and mathematical modeling are shown in Figs. 5.3 and 5.4.

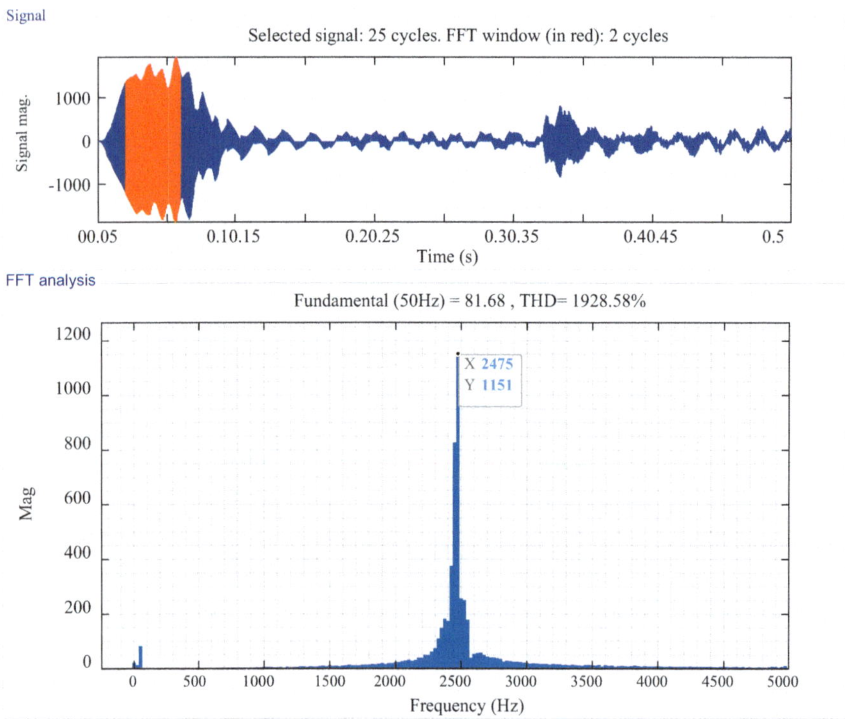

Fig. 5.6 Inverter output current waveform and FFT analysis diagram with $L_g = 10$mH and $T_2 = 0$

The transfer function of Lead compensation is expressed:

$$T_2(s) = k_d \frac{sT_d + 1}{s\alpha T_d + 1}. \tag{5.8}$$

In order to see the impact of parameter changes on the Lead compensation module more intuitively, the bode diagrams of parameter changes are drawn as shown in Fig. 5.7. Figure 5.7a shows the bode diagram of k_d changes with $\alpha = 0.1325$ and $T_d = 1/4000\pi$; Fig. 5.7b shows the bode diagram of α changes with $k_d = 0.1$ and $T_d = 1/4000\pi$; Fig. 5.7c shows the bode diagram of T_d changes with $k_d = 0.1$ and $\alpha = 0.1325$.

From Fig. 5.7a, it can be seen that: (1) the amplitude-frequency characteristic curve of the Lead link moves upward and the phase-frequency characteristic curve remains unchanged when α and T_d remain unchanged and k_d increases. (2) The amplitude-frequency gain is k_d at low frequencies with $s \to 0$ and the amplitude-frequency gain is k_d / α at high frequencies with $s \to \infty$.

From Fig. 5.7b, it can be seen that: when k_d and T_d remain unchanged and α increases, the amplitude-frequency characteristic curve of the Lead link is significantly shifted down in the high-frequency band, the compensation angle of the phase-frequency characteristic curve is reduced, and the differential effect on the output signal is weakened.

From Fig. 5.7c, it can be seen that: when k_d and α remain unchanged and T_d decreases, the frequency at the maximum compensation angle of the Lead link increases, but the high-frequency gain and the maximum compensation angle remain unchanged.

In order to determine the numerical relationship between compensation angle, compensation frequency, and Lead link parameters, let $s = j\omega$, it can be obtained from formula (5.8):

$$T_2(j\omega) = k_d \frac{j\omega T_d + 1}{j\omega\alpha T_d + 1}. \tag{5.9}$$

The phase-frequency expression of $T_2(j\omega)$ is

$$\varphi_p(\omega) = \arctan\left(\frac{\omega T_d - \alpha\omega T_d}{1 + \alpha\omega^2 T_d^2}\right). \tag{5.10}$$

Let $d\varphi_p / d\omega = 0$, the angular frequency at the maximum phase compensation angle can be obtained as

$$\omega_m = \frac{1}{T_d\sqrt{\alpha}}. \tag{5.11}$$

Then, by substituting ω_m into formula (5.11), the maximum leading angle can be obtained as

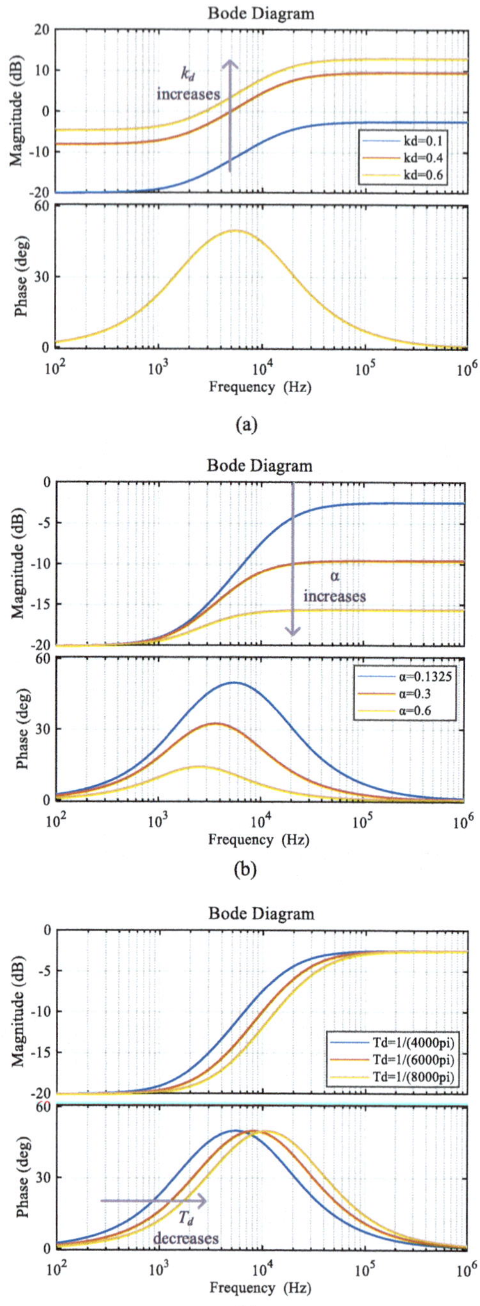

Fig. 5.7 Bode diagram of the effect of parameters changes with Lead compensation. **a** Bode diagram of k_d changes with $\alpha = 0.1325$ and $T_d = 1/4000\pi$. **b** Bode diagram of k_d changes with $\alpha = 0.1$ and $T_d = 1/4000\pi$. **c** Bode diagram of T_d changes with $k_d = 0.1$ and $\alpha = 0.1325$

$$\varphi_{\max} = \sin^{-1}\frac{1-\alpha}{1+\alpha}. \tag{5.12}$$

In order to keep the amplitude-frequency constant at the maximum phase compensation angle while compensating the phase margin of the system, k_d should be used to adjust the gain so that $|T_2(j\omega_m)| = 1$,which can be obtained from (5.9):

$$k_d = \frac{1+\frac{1}{\alpha}}{\sqrt{2+\alpha+\frac{1}{\alpha}}}. \tag{5.13}$$

The above analysis indicates that the parameter k_d of Lead compensation affects the compensation gain, α affects the maximum compensation angle, and T_d and α together affect the compensation frequency.

Capacitor voltage Lead compensation feedforward method simulation verification

Figure 5.8 shows the open-loop bode diagram of the current inner loop when the capacitor voltage Lead compensation feedforward is added. The parameters of the three-phase grid-connected inverter in Fig. 5.8 are the same as those in Fig. 5.5. The specific parameters of the capacitor voltage Lead compensation feedforward method are set as the compensation gain $k_d = 0.4$, the maximum compensation angle $\varphi_{\max} = 50°$, and the compensation frequency being 6000 Hz, then, the Lead compensation efficiency α can be obtained as

$$\alpha = \frac{1-\sin\varphi_{\max}}{1+\sin\varphi_{\max}} = 0.1325. \tag{5.14}$$

The bode diagram in Fig. 5.8 shows that when the capacitor voltage Lead compensation feedforward is added, the GM is 27.5 dB, the PM is 10.1 deg, and the GM

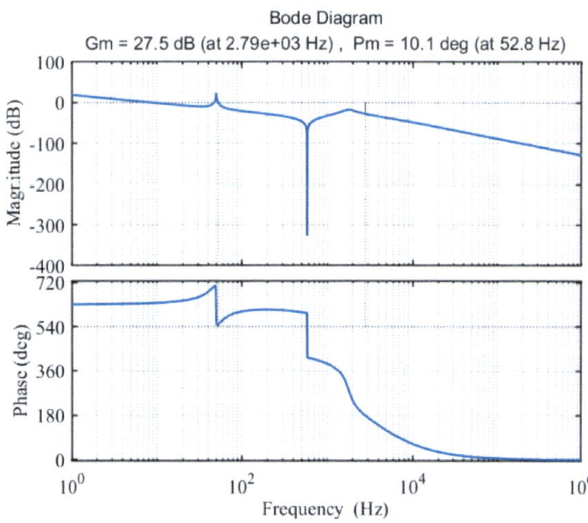

Fig. 5.8 Open loop bode diagram of the current inner loop when the capacitor voltage Lead compensation feedforward is added

and PM of the three-phase grid-connected inverter system are both greater than 0. At this time, the three-phase grid-connected inverter system can operate stably.

Figure 5.9 shows the simulation results of the LCL three-phase grid-connected inverter at the rated load in MATLAB/Simulink with capacitor voltage Lead compensation feedforward. Figure 5.9a shows the waveform of three-phase grid-connected current i_{L2}, and Fig. 5.9b shows the FFT analysis results of i_{L2}. From the simulation waveforms in Fig. 5.9, we can see that: (1) The 20 kW three-phase grid-connected inverter runs stably, and the amplitude of grid-connected current i_{L2} is about 42.8 A, which is close to the theoretical value. (2) The output current waveform of the three-phase grid-connected inverter has a good sinusoidal effect, and THD is only 1.18%. The simulated waveforms from Fig. 5.9 coincide with the results of the bode diagram analyzed in the previous section, thus verifying the effectiveness of the capacitor voltage Lead compensation feedforward strategy to suppress the LCL resonance under the weak power grid of the three-phase grid-connected inverter.

In practice, since the voltage of the filter inductor L_2 is very small, it is approximately considered that the PCC voltage is equal to the voltage of the filter capacitor, i.e.,$u_{pcc} \approx u_c$.

5.2.4 Reactive Power Compensation Method for PCC Voltage Control

Operation analysis of PV power plant under unit power factor

In this article, the PV grid-connected inverter adopts current control mode, so the entire PV farm is approximated as a fundamental frequency current source, and a simplified circuit of the entire PV grid-connected system is obtained as shown in Fig. 5.10, where i_{pv} is the grid-connected current injected by the PV power plant, i_{L2}, u_{pcc}, u_g and Z_g are the PCC voltage, grid voltage, and grid impedance, respectively.

According to formula (5.1), the grid impedance Z_g can be calculated as follows:

$$\left|Z_g\right| = \frac{U_g^2}{P_{pv\_rated} \cdot SCR}, \tag{5.15}$$

where P_{pv_rated} is the rated power of the PV power station and U_g is the root mean square (RMS) value of u_g.

When the PV plant operates under a unity power factor, the worst situation of the power grid is considered to be pure inductive, and the vector relationship between the PV plant and the grid is obtained as shown in Fig. 5.11, where U_{pcc}, $U_{g,}$ and I_{pv} are the vector forms of u_{pcc}, u_g, and i_{pv}, respectively, the value Z$_g$ is $Z_g = jX_g$.

According to Fig. 5.11, the formula of the U_{pcc} is as follows:

$$U_{pcc} = \sqrt{U_g^2 - \left(X_g \cdot I_{pv}\right)^2}. \tag{5.16}$$

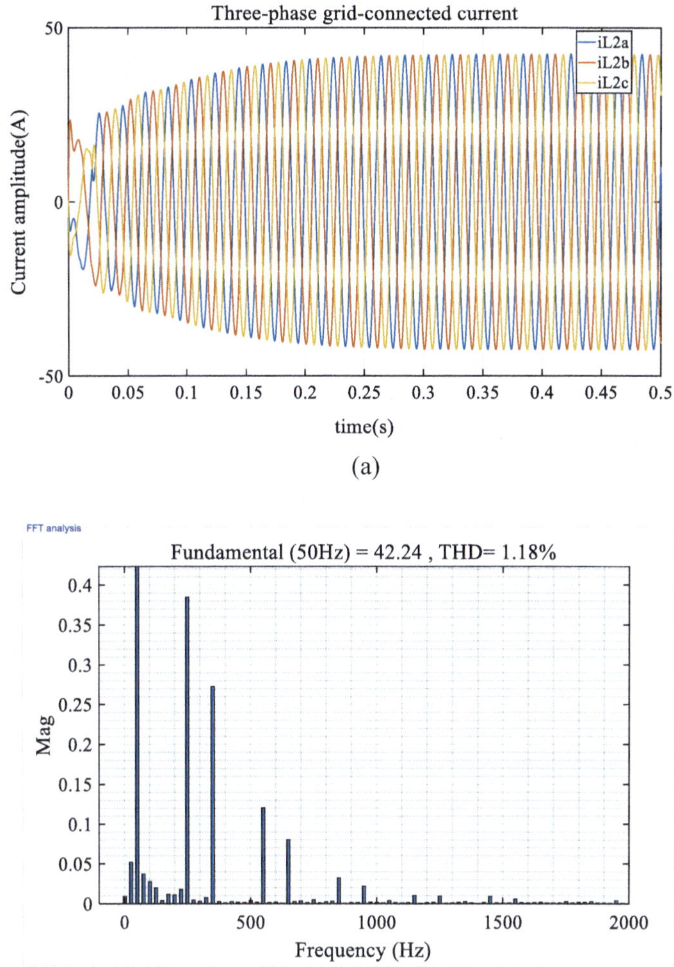

Fig. 5.9 Grid-connected current i_{L2} and FFT analysis waveforms with $L_g = 10\text{mH}$ with capacitor voltage Lead compensation feedforward. **a** Three-phase grid-connected current i_{L2} waveforms. **b** FFT analysis of i_{L2}

Fig. 5.10 Simplified circuit of PV grid-connected system

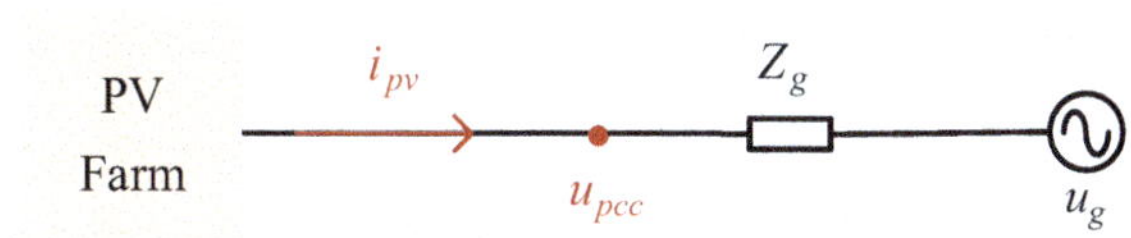

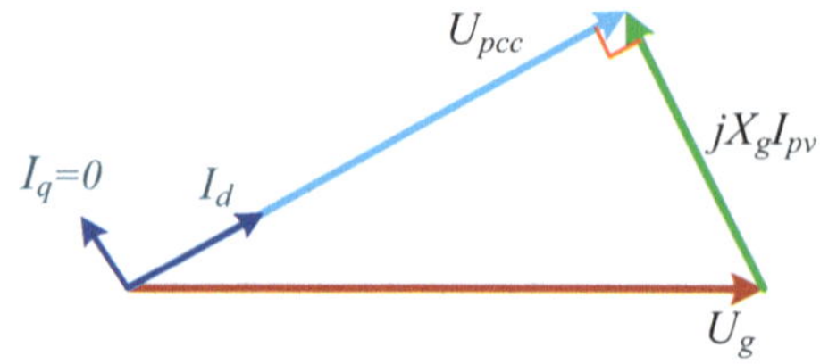

Fig. 5.11 Vector relationship between the PV power station and the power grid with PF = 1

The active power injected by PV power plant is

$$P_{pv} = U_{pcc} \cdot I_{pv} = \sqrt{U_g^2 - \left(X_g \cdot I_{pv}\right)^2} \cdot I_{pv}. \tag{5.17}$$

The maximum value of active power can be obtained with satisfying as

$$\frac{\partial P_{pv}}{\partial I_{pv}} = 0. \tag{5.18}$$

In terms of (5.17) and (5.18), the maximum value of active power is obtained as

$$P_{pv\_max} = X_g I_{pv}^2 = \frac{U_g^2}{2X_g}. \tag{5.19}$$

Combining formula (5.15) and (5.19) with $P_{pv\_max} \geq P_{pv\_rated}$, the minimum SCR of the grid-connected inverter is $SCR_{\min} = 2$ with the PF = 1.

Operation analysis of PV power plant under variable power factors

To achieve stable operation of PV grid-connected inverters under more weak power grid conditions, the common method is to install large-capacity flexible AC transmission system facts (FACTS) to provide reactive power compensation or eliminate voltage drop in transmission lines. However, to reshape the system impedance under extremely weak power grid conditions, the reactive power needed to be provided by FACTS devices will be large, making the installation cost of FACTS devices high. In order to save equipment volume and cost, reactive power can also be provided by the PV power plant itself. At this time, the PV power plant needs to operate under variable PFs. The specific vector relationship diagram is shown in Fig. 5.12, where the phase angle is φ.

where $I_d = \cos\varphi$ is consistent with U_{pcc} direction, $I_q = \sin\varphi$ is perpendicular to U_{pcc} direction.

As shown in Fig. 5.12, the active power injected by the PV power plant is as follows:

$$P_{pv} = U_{pcc} \cdot I_{pv} = \left(\sqrt{U_g^2 - \left(X_g \cdot I_{pv} \cos\varphi\right)^2} + X_g \cdot I_{pv} \sin\varphi\right) \cdot I_{pv} \cos\varphi. \tag{5.20}$$

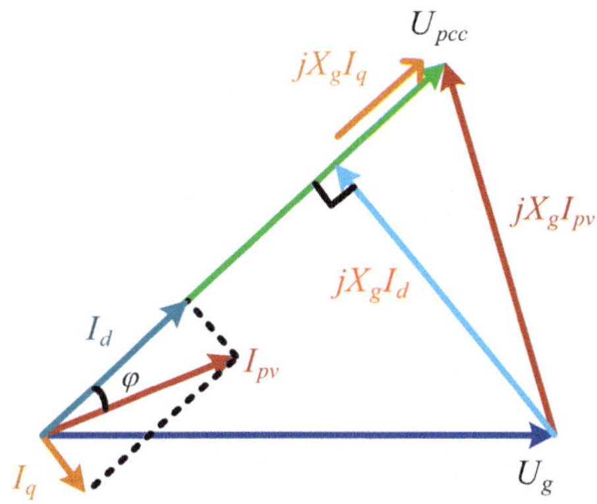

Fig. 5.12 Vector relationship between the PV power plant and the power grid when PF is variable

In order to transmit the rated power to the power grid, some factors such as loss should be considered. Combined with formula (5.20), the following formula can be obtained:

$$\left(\sqrt{1-\left(\frac{X_g}{U_g}\cdot I_{pv}\cos\varphi\right)^2}+\frac{X_g}{U_g}\cdot I_{pv}\sin\varphi\right)\cdot\frac{I_{pv}}{I_{pv\_rated}}\cos\varphi>1, \tag{5.21}$$

where the rated power P_{pv_rated} is expressed as

$$P_{pv\_rated}=I_{pv\_rated}\cdot U_g. \tag{5.22}$$

According to formula (5.21), the ratio of X_g to U_g is

$$\frac{X_g}{U_g}=\frac{U_g}{P_{pv\_rated}\cdot SCR}=\frac{1}{I_{pv\_rated}\cdot SCR}. \tag{5.23}$$

Combined with (5.22) and (5.23), it can be obtained as

$$\left(\sqrt{1-\left(\frac{I_{pv(pu)}\cos\varphi}{SCR}\right)^2}+\frac{I_{pv(pu)}\sin\varphi}{SCR}\right)\cdot I_{pv(pu)}\cos\varphi>1, \tag{5.24}$$

where $I_{pv(pu)}=I_{pv}/I_{pv\_rated}$ is the per unit value of the current injected into the power grid of the PV power plant.

In terms of (5.21), the minimum SCR can be obtained as follows:

$$SCR_{\min}=\frac{I_{pv(pu)}^2\sin\varphi\cos\varphi-\left(I_{pv(pu)}\cos\varphi\right)^2\sqrt{I_{pv(pu)}^2-1}}{1-\left(I_{pv(pu)}\cos\varphi\right)^2}. \tag{5.25}$$

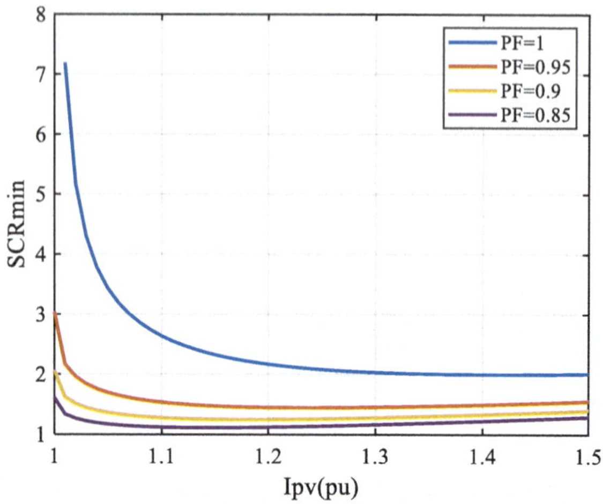

Fig. 5.13 Relationship between SCR_{min} and $I_{pv(pu)}$ with different PFs

Figure 5.13 shows the relationship between SCR_{min} and the per unit current ($I_{pv(pu)}$) injected into the power grid by the PV power plant under different PF conditions. From Fig. 5.13, it can be seen that SCR_{min} correspondingly decreases when PF decreases, and the PV power plant can operate at a smaller minimum SCR. Although smaller PF can enable PV power plants to operate under weaker power grids, generally speaking, the value of PF will not be less than 0.9. When the PF value is too small, it will have adverse effects such as insufficient utilization of power equipment capacity and excessive line losses.

5.2.5 Simulation Verification of Reactive Power Compensation Method Controlled by PCC Voltage

From Fig. 5.11, it can be seen that the grid voltage U_g remains constant and the value of U_{pcc} will decrease, resulting in a phenomenon of voltage drop at PCC if the grid impedance continues to increase with SCR < 2 and jX_gI_{pv} increasing. As shown in Fig. 5.12, SCR_{min} can gradually decrease when $0.85 < PF < 1$. At this time, the reactive power of the PV power plant is compensated through the *q*-axis current, and the stability of the U_{pcc} is controlled through jX_gI_q.

The PCC voltage control method is applied to the control of the grid-connected inverter in the PV power plant. The outer loop control diagram of the PV grid-connected inverter is shown in Fig. 5.14. Compare the DC voltage reference u_{dcref} and the actual sampled DC voltage u_{dc} to obtain the error signal, and the reference current i_{dref} is obtained by PI controller. Similarly, compare the PCC reference voltage $u_{pccdref}$ and the actual sampled PCC voltage u_{pccd} to obtain the error signal, and the reference current i_{qref} is obtained by PI controller. Then, the reference currents $i_{\alpha ref}$ and $i_{\beta ref}$ are obtained by inverse Park transformation of i_{dref} and i_{qref}. The inverse

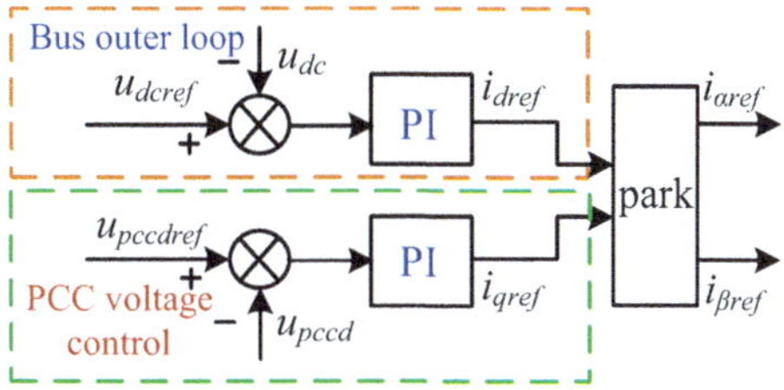

Fig. 5.14 Outer loop control block diagram of PV grid-connected inverters

Park transformation formula is as follows:

$$\begin{bmatrix} x_\alpha \\ x_\beta \end{bmatrix} = \begin{bmatrix} \cos\theta & -\sin\theta \\ \sin\theta & \cos\theta \end{bmatrix} \begin{bmatrix} x_d \\ x_q \end{bmatrix}, \quad (5.26)$$

where θ is the angle obtained by the phase-locked loop.

Three-phase grid-connected current i_{L2x}(x = a,b,c) and the active and reactive powers are shown in Fig. 5.15 without PCC voltage control under L_g = 10mH and SCR = 2.312. As can be seen from the simulation results in Fig. 5.15, when SCR is greater than 2, the system can run stably without the PCC voltage control for reactive power compensation, and the output active power of the three-phase grid-connected inverter is 20 kW and reactive power is 0.

Similarly, three-phase grid-connected current i_{L2x}(x = a,b,c) and the active and reactive powers are shown in Fig. 5.16 without PCC voltage control under L_g = 20mH and SCR = 1.16. It can be seen from Fig. 5.16 that when SCR is less than 2, the three-phase grid-connected inverter system is unstable without the PCC voltage control for reactive power compensation.

Three-phase grid-connected current i_{L2x}(x = a,b,c) and the active and reactive powers are shown in Fig. 5.17 with PCC voltage control under L_g = 20mH and SCR = 1.16. As can be seen from the simulation results shown in Fig. 5.17, the three-phase grid-connected inverter can run stably with the PCC voltage control under SCR less than 2, and the output active power and reactive power of the three-phase grid-connected inverter are 20 kW and 10 kW, respectively. According to the simulation results shown in Fig. 5.17, it can be seen that the three-phase grid-connected inverter with PCC voltage control has stable performance under weak power grid conditions, thus verifying the effectiveness of the PCC voltage control.

Summary of control strategy of three-phase grid-connected inverter under weak power grid

First, this section analyzes the control structure and mathematical model of LCL three-phase grid-connected inverter with inverter current feedback control. Then, the capacitor voltage Lead compensation feedforward method and the PCC voltage control reactive power compensation method are used to solve the LCL resonance problem and the PCC voltage drop problem when the three-phase grid-connected inverter operates under a weak power grid. The effectiveness of the proposed scheme

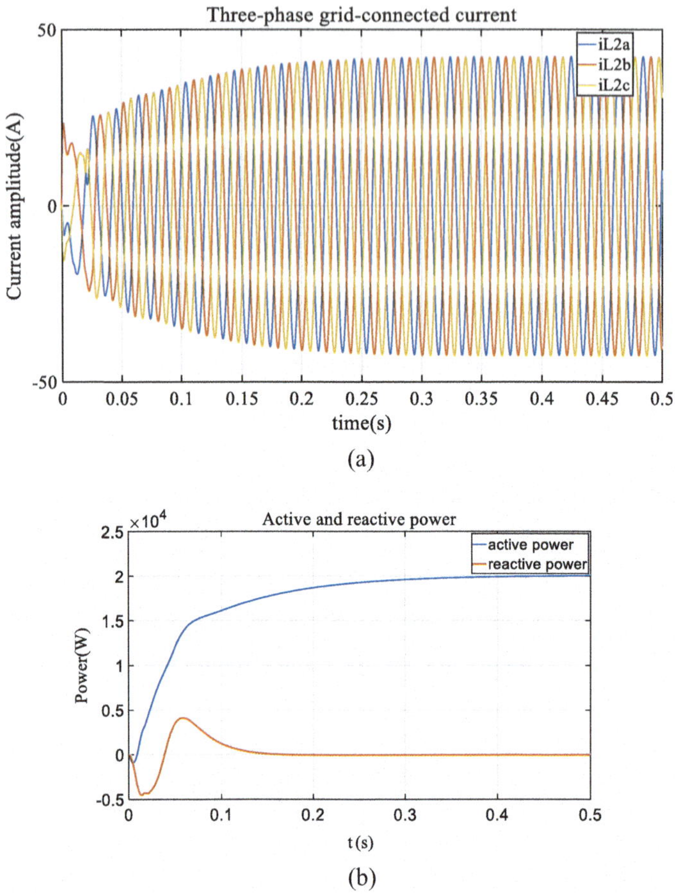

Fig. 5.15 Simulation waveforms without PCC voltage control under $L_g = 10$mH and SCR = 2.312. **a** Three-phase grid-connected output currents. **b** Active and reactive power

is verified through comparative simulation experiments in MATLAB/Simulink, achieving the goal of stable operation of PV grid-connected inverters under weak power grid conditions of SCR = 1.16.

5.3 Research Status of Grid Impedance Detection Methods

When the three-phase grid-connected inverter operates under weak grid conditions, in order to effectively control the three-phase grid-connected inverter, it is necessary to detect the grid impedance. Existing grid impedance detection methods for grid-connected inverters are mainly divided into active detection and passive detection. Passive detection methods detect grid impedance based on the system's inherent

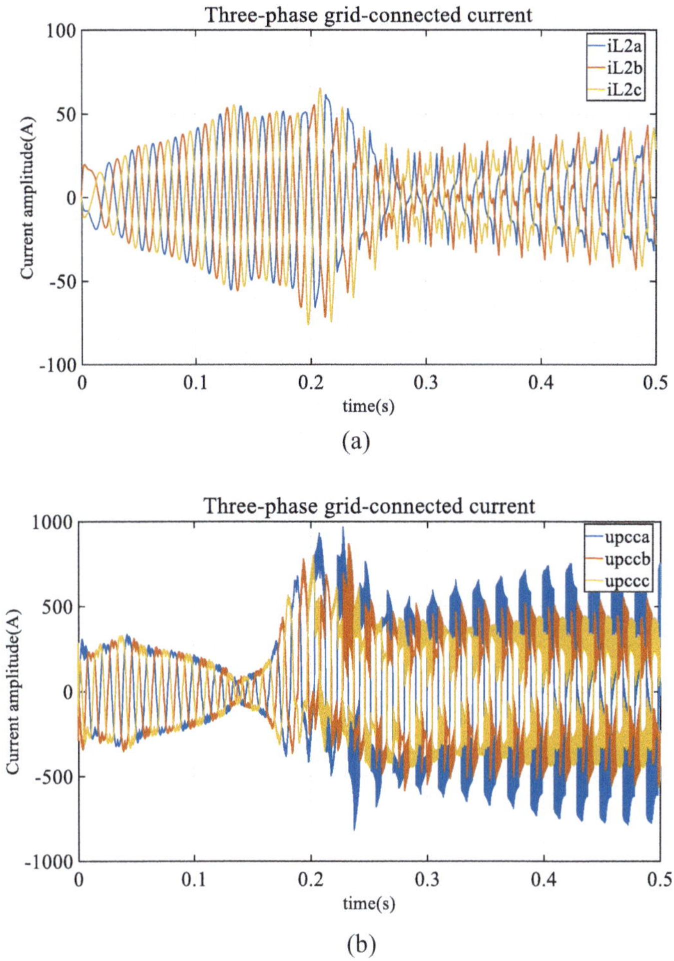

Fig. 5.16 Simulation waveforms without PCC voltage control under $L_g = 20\text{mH}$ and SCR = 1.16. **a** Three-phase grid-connected output currents. **b** Three-phase grid-connected voltages

characteristics without the need to inject disturbances into the system. Passive detection methods mainly include the recursive least squares method, successive iterative approximation method, and extended Kalman filter method, etc. [16–18]. In recent years, various new grid impedance detection methods have been proposed, such as the method based on ripple analysis and the structural characteristics of the DC microgrid system to detect the grid impedance, and the method using channel frequency response (CFR) to track grid impedance, etc. [19, 20]. The advantages of these methods are that the stable operation of grid-connected inverter and the quality of output current waveform are not affected. However, the drawback is excessive dependence on background distortion signals. For example, when the signal to noise

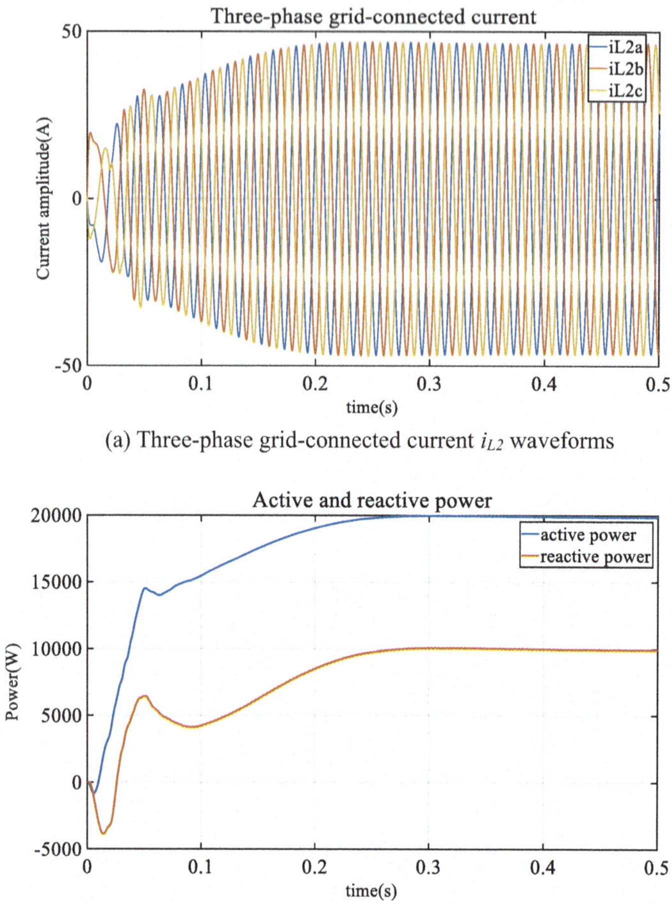

(a) Three-phase grid-connected current i_{L2} waveforms

(b) Active and reactive power

Fig. 5.17 Simulation waveforms with PCC voltage control under $L_g = 20\text{mH}$ and SCR = 1.16. **a** Three-phase grid-connected output currents. **b** Active and reactive power

ratio (SNR) of the grid-connected inverter system is low, the detection accuracy will be low. Besides, passive detection methods are generally computationally complex, difficult to determine parameters, and difficult to implement in practical systems.

To overcome the issues associated with passive detection methods, active grid impedance detection methods have been proposed by scholars at home and abroad. The active detection methods need to inject disturbance into the system and extract corresponding disturbance signals at the point of common coupling (PCC) to calculate the grid impedance. The disturbance injection types for the active detection methods mainly include exciting LCL resonances, PQ power disturbance injection, broadband pulse injection, single harmonic injection, etc. [20–22]. In [20], the disturbance variations of PQ power were used to detect the grid impedance. However, this method is only applicable to PQ-controlled grid-connected inverter systems. In [21], broadband pulse injection was used to detect the grid impedance. The advantage of this method is that it can detect the grid impedance within a wider spectral range, as opposed to a certain frequency point, but the lower detection accuracy and heavier computational burden are its main drawbacks. Currently, the most commonly used disturbance injection method is the single harmonic injection. To reduce the impact of disturbance injection on the system, in [22], research was conducted on the amplitude, duration, and frequency of the injected non-characteristic harmonic.

The above active detection methods mostly use Discrete Fourier Transform (DFT) to extract harmonics. However, in practical engineering, mutual interference between spectrums can affect detection accuracy. In order to overcome the limitations of DFT, a transient grid impedance detection method based on continuous wavelet transform (CWT) was proposed in [23]. Nevertheless, due to the complexity of CWT and the high redundancy of information, this method is difficult to apply in practice. In recent years, the complex coefficient filter (CCF) method has gradually been applied in electrical engineering and power systems.

In [24], the combination of CCF and phase lock loop (PLL) can accurately and rapidly extract the positive and negative sequence components of the fundamental wave and other harmonic components in the case of unbalanced grid voltage. Reference [25] proposed a dead-time compensation method based on a modified multiple CCF to optimize the control performance of permanent magnet synchronous motors. Reference [26] applied the CCF method to grid impedance detection to extract non-characteristic harmonic components of grid-connected signals, which can detect accurately in both ideal and unbalanced grids. In this chapter, an active impedance detection method based on single harmonic injection and complex coefficient filter is proposed.

5.4 Grid Impedance Detection Method Based on Complex Coefficient Filter

5.4.1 Grid Impedance Detection Method Based on Non-Characteristic Harmonic Injection

The simplified model of the grid-connected inverter system is shown in Fig. 5.18, where u_i is the inverter output voltage.

By injecting non-characteristic current harmonic disturbances at the given current, the corresponding harmonic response signals $u_{pcc}(h)$ and $i_{L2}(h)$ at the point of common coupling are extracted for grid impedance detection, where h is the ratio of the non-characteristic harmonic frequency to the fundamental frequency (50 Hz). The specific detection formula is as follows:

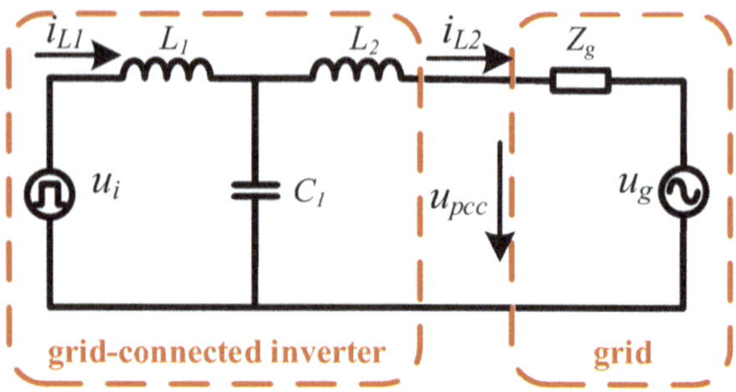

Fig. 5.18 Simplified model of the grid-connected inverter system

$$\hat{Z_g}(h) = \frac{\Delta \hat{u}(h)}{\Delta \hat{i}(h)} = \frac{\hat{u}_{pcc}(h) - \hat{u_g}(h)}{\hat{i}_{L2}(h)}, \tag{5.27}$$

where $\hat{u_g}(h)$ is the non-characteristic voltage harmonic signal in the power grid, which can be ignored due to its small value compared with $\hat{u}_{pcc}(h)$. Consequently, (5.27) can be simplified as

$$\hat{Z_g}(h) = \frac{\hat{u}_{pcc}(h)}{\hat{i}_{L2}(h)} = R_g + j\omega L_g. \tag{5.28}$$

The correctness of the proposed method has been verified under ideal grid conditions. However, when the grid is contaminated, the estimated grid impedance will fluctuate. This problem can be solved by appropriately adjusting the amplitude and frequency of the injected harmonics.

5.4.2 *Complex Coefficient Filter Harmonic Extraction Method*

Introduction to the first-order complex filter.

In this chapter, the complex coefficient filter method is applied to extract harmonics from the sampled harmonic response signals $u_{pcc}(h)$ and $i_{L2}(h)$. The formulas for extracting positive and negative sequence components of the fundamental wave by the first-order complex coefficient filter are expressed as

$$H_{CCF+}(s) = \frac{\omega_c}{s - j\omega_0 + \omega_c} \tag{5.29}$$

$$H_{CCF-}(s) = \frac{\omega_c}{s + j\omega_0 + \omega_c}, \tag{5.30}$$

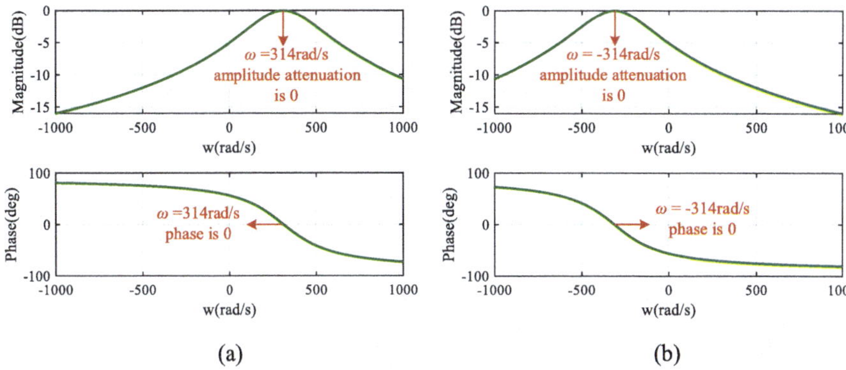

Fig. 5.19 Bode diagrams of the first-order complex coefficient filter used to extract the positive and negative sequence components of the fundamental wave. **a** Extracting positive sequence component. **b** Extracting negative sequence component

where the fundamental wave angular frequency ω_0 is 314.16 rad/s, and the cutoff frequency ω_c is 221 rad/s.

The bode diagrams of the first-order complex coefficient filter used to extract the positive and negative sequence components of the fundamental wave are shown in Fig. 5.19, where Fig. 5.19a is the case of extracting the positive sequence component and Fig. 5.19b is the case of extracting the negative sequence component. From the bode diagrams in Fig. 5.19, it can be seen that for the first-order complex coefficient filters, the amplitude attenuation is 0 and the phase is 0 at the positive or negative sequence component of the fundamental wave. Meanwhile, the amplitude has significant attenuation at other frequencies. Compared with the second-order band pass filter, the complex coefficient filter can combine the in-phase and quadrature signals for observation, suppress the fundamental frequency negative sequence component, and retain the fundamental frequency positive sequence component.

Complex coefficient filter harmonic extraction method

Based on the requirements of harmonic extraction for grid impedance detection, it is necessary to suppress the positive and negative sequence components of the fundamental frequency signals (there is a negative sequence component when the three-phase grid voltage is unbalanced) and extract the injected non-characteristic harmonic signals. In order to meet this requirement, the structure diagram of the complex coefficient filter adopted in this chapter is shown in Fig. 5.20, where x represents the grid-connected voltage and current signal extracted for grid impedance detection. The control structure of a complex coefficient filter (CCF) includes three parts: the block A extracts non-characteristic signals $\mathrm{x}^{\mathrm{h}}_{\alpha\beta}$ $\mathrm{x}^{\mathrm{h}}_{\alpha\beta}$ $\mathrm{x}^{\mathrm{h}}_{\alpha\beta}\mathrm{x}^{\mathrm{h}}_{\alpha\beta}$, the block B extracts the fundamental frequency positive sequence component of the signals $x^{+1}_{\alpha\beta}$ $\mathrm{x}^{+1}_{\alpha\beta}\mathrm{x}^{+1}_{\alpha\beta}$, while the block C extracts the fundamental frequency negative sequence component of the signals $x^{-1}_{\alpha\beta}$ $\mathrm{x}^{-1}_{\alpha\beta}$ $\mathrm{x}^{-1}_{\alpha\beta}$ [27].

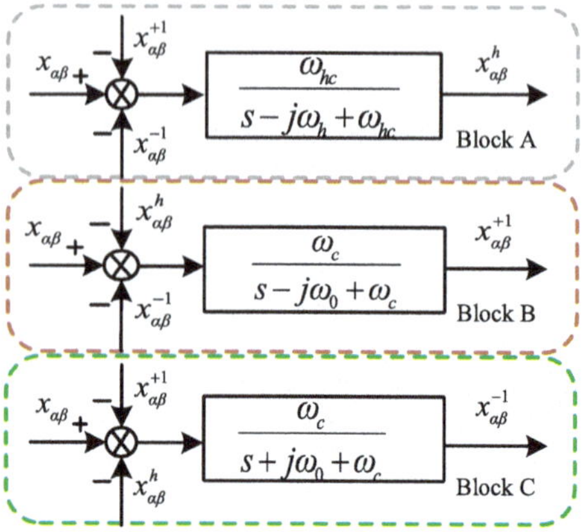

Fig. 5.20 Structure diagram of the complex coefficient filter harmonic extraction

According to Fig. 5.20, the transfer function for extracting non-characteristic harmonic signals can be calculated as

$$H_{CCF}(s) = \frac{\omega_{hc}\left(s^2 + \omega_0^2\right)}{s^3 + As^2 + Bs + C\omega_0^2}, \tag{5.31}$$

where A = $2\omega_c$-$j\omega_h + \omega_{hc}$, B = ω_0^2 ω_0^2 ω_0^2-$j2\omega_h\omega_c$, C = ω_{hc}-$j\omega_h$. ω_0 is the fundamental wave angular frequency, ω_c is the fundamental wave cutoff angular frequency, ω_h is the non-characteristic harmonic angular frequency, ω_{hc} is the non-characteristic harmonic cutoff angular frequency.

Figure 5.21 shows the amplitude-frequency characteristic curve and closed-loop pole distribution diagram of the complex coefficient filter (CCF) with $\omega_0 =$ 314.16 rad/s, $\omega_c = 221$ rad/s and $\omega_h = 1728$ rad/s when ω_{hc} is different. It can be found from Fig. 5.21 that: (1) The smaller ω_{hc} is, the better the filtering performance

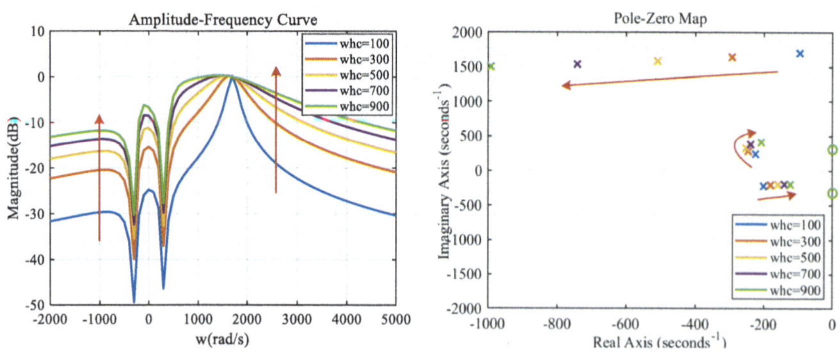

Fig. 5.21 CCF amplitude-frequency characteristic curve and closed-loop pole distribution diagram when ω_{hc} is different

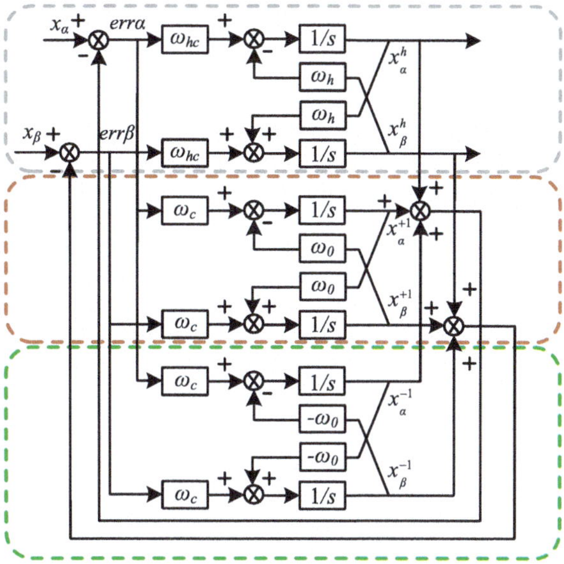

Fig. 5.22 Block diagram of CCF implementation in αβ coordinate system

of the complex coefficient filter (CCF) model is and the higher the detection accuracy is. However, two groups of closed-loop poles will move closer to the imaginary axis, which will make the dynamic performance of the system worse, so the selection of parameters needs to consider both detection accuracy and dynamic performance comprehensively. (2) The amplitude-frequency characteristic curve of CCF also shows that this model can suppress the positive and negative sequence components of fundamental wave in the signals and extract non-characteristic harmonic components, which meets the requirements of grid impedance detection.

In order to change the complex number j in Fig. 5.20 into a code realizable form, the above principle is implemented in αβ coordinate system, where j represents the phase difference between α and β. The transformation process is depicted in Fig. 5.22.

In a practical digital implementation, the control algorithm will be implemented and executed in a digital signal processor (DSP). Therefore, the system models and control signals need to be discretized by the Tustin method. Taking the α phase non-characteristic signal as an example, the discretization extraction formula is as follows:

$$\left[err\alpha(k)\omega_{hc} - x_\beta^h(k)\omega_h\right]\frac{T_s}{2}\frac{1+z^{-1}}{1-z^{-1}} = x_\alpha^h(k) \tag{5.32}$$

$$x_\alpha^h(k) = x_\alpha^h(k-1) + \frac{T_s}{2}\left((err\alpha(k) + err\alpha(k-1))\omega_{hc} - \left(x_\beta^h(k) + x_\beta^h(k-1)\right)\omega_h\right). \tag{5.33}$$

5.4.3 Grid Impedance Detection Method Based on CCF and Full-Order Capacitor Current Observer

Full-order capacitor current observer

According to the above analysis, non-characteristic grid-connected voltage signal $u_{pcc}(h)$ and grid-connected current signal $i_{L2}(h)$ need to be extracted for grid impedance detection. In order to reduce the use of current sensors in grid-connected inverters, a full-order capacitor current observer is used to estimate the capacitor current. Subsequently, the grid current is calculated by the sampled inverter current and the observed capacitor current.

Since the voltage of the filter inductor L_2 is particularly small compared with the grid-connection voltage, it can be approximated that the sampled common coupling point voltage is equal to the capacitor voltage, namely $u_{pcc} \approx u_c$, the capacitor current is estimated according to the grid-connected voltage. Simultaneously, since the sampling period $T_s = 1/f_s$ is very small compared with the fundamental wave period, it can be assumed that the grid-connected current between two adjacent sampling moments is constant, namely $\mathrm{di_{L2}/dt} = 0$ $\mathrm{di_{L2}/dt} = 0$ $di_{L2}/dt = 0$.

According to the control block diagram in Fig. 5.3, the following formulas can be obtained:

$$\frac{du_{pcc}}{dt} = \frac{1}{C_1} i_c \tag{5.34}$$

$$\frac{di_c}{dt} = \frac{di_{L1}}{dt} - \frac{di_{L2}}{dt} = \frac{1}{L_1}\left(u_i - u_{pcc}\right), \tag{5.35}$$

where u_i is the inverter output voltage, which is replaced by the modulated wave.

The full-order capacitor current observer based on this model structure can be mathematically described as

$$\frac{d\hat{u}_{pcc}}{dt} = \frac{1}{C_1}\hat{i}_c + k_1\left(u_{pcc} - \hat{u}_{pcc}\right). \tag{5.36}$$

$$\frac{d\hat{i}_c}{dt} = \frac{1}{L_1}\left(u_i - \hat{u}_{pcc}\right) + k_2\left(u_{pcc} - \hat{u}_{pcc}\right), \tag{5.37}$$

where ∧ represents the observed value, and k_1 and k_2 are the observation coefficients of capacitor current. By discretizing formulas (5.36) and (5.37), the observed capacitor current can be obtained as

$$\hat{u}_{pcc}(k+1) = \hat{u}_{pcc}(k) + T_s\left(\frac{1}{C_1}\hat{i}_c(k) + k_1 e(k)\right) \tag{5.38}$$

$$\hat{i}_c(k+1) = \hat{i}_c(k) + T_s\left(\frac{1}{L_1}\left(u_i(k) - \hat{u}_{pcc}(k)\right) + k_2 e(k)\right), \tag{5.39}$$

where $e(k) = u_{pcc} - \hat{u}_{pcc}$ is the voltage observation error, $u_i(k)$ is the modulated wave at the previous time, and the above formulas are applicable in the αβ coordinate system.

Determination of capacitor current observation coefficient

In order to determine the values of the observation coefficients k_1 and k_2, the stability of the observer is analyzed by the pole placement method. From formulas (5.38) and (5.39), it can be concluded that:

$$\underbrace{\begin{bmatrix} \hat{u}_{pcc}(k+1) \\ \hat{i}_c(k+1) \end{bmatrix}}_{\hat{x}(k+1)} = \underbrace{\begin{bmatrix} 1 & \frac{T_s}{C_1} \\ -\frac{T_s}{L_1} & 1 \end{bmatrix}}_{A} \underbrace{\begin{bmatrix} \hat{u}_{pcc}(k) \\ \hat{i}_c(k) \end{bmatrix}}_{\hat{x}(k)} + \underbrace{\begin{bmatrix} 0 \\ \frac{T_s}{L_1} \end{bmatrix}}_{B} u_i(k) + \underbrace{\begin{bmatrix} k_1 T_s & 0 \\ k_2 T_s & 0 \end{bmatrix}}_{C} \begin{bmatrix} x(k) - \hat{x}(k) \end{bmatrix}$$

$$= (A - C)\,\hat{x}(k) + Bu_i(k) + Cx(k), \tag{5.40}$$

where A, B, and C are the coefficient matrices of the capacitor current observer $x(k) = \begin{bmatrix} \hat{u}_{pcc}(k) & \hat{i}_c(k) \end{bmatrix}^T$.

To ensure that the observer is in a stable state in the discrete domain, all eigenvalues of the observer coefficient matrix (AC) in the z-plane are required to be in the unit circle, and the characteristic polynomial expression is as follows:

$$\det(zI - A + C) = z^2 + (k_1 T_s - 2)z + 1 - k_1 T_s + \frac{T_s}{C_1}\left(k_2 T_s + \frac{T_s}{L_1}\right). \tag{41}$$

Furthermore, in order to achieve a fast dynamic response of the proposed full-order capacitor current observer, the two eigenvalues are set to zero, namely,

$$\det(zI - A + C) = z^?. \tag{42}$$

Comparing formulas (5.38) and (5.39), the observation coefficients k_1 and k_2 of capacitor current are

$$\begin{cases} k_1 T_s - 2 = 0 \\ 1 - k_1 T_s + \frac{T_s}{C_1}\left(k_2 T_s + \frac{T_s}{L_1}\right) = 0 \end{cases} \tag{43}$$

$$\begin{cases} k_1 = \dfrac{2}{T_s} \\ k_2 = \dfrac{C_1}{T_s^2} - \dfrac{1}{L_1} \end{cases}. \tag{44}$$

The observed value of capacitor current can be obtained by substituting the observation coefficients into formulas (5.38) and (5.39).

Grid impedance detection method combining CCF and full-order capacitor current observer

In order to reduce the cost and improve the reliability of inverter, a new grid impedance detection method combining a complex coefficient filter (CCF) and full-order capacitor current observer is proposed. Compared with the conventional CCF impedance detection algorithm, the proposed method reduces the number of current sensors and can accurately detect the grid impedance value. The overall control block diagram of the system is shown in Fig. 5.23. This chapter conducts research on a T-type three-level three-phase grid-connected inverter with inverter current feedback control and LCL filter. The control system includes inverter control strategy and grid impedance detection, where the grid impedance detection is divided into harmonic injection and harmonic extraction [27].

Firstly, non-characteristic harmonic signals are injected at the reference current of the three-phase grid-connected inverter. Secondly, the grid-connected current can be calculated through the sampled inverter output current and the capacitor current estimated by the full-order capacitor current observer. Then, the calculated grid-connected current signal and the sampled common coupling point voltage signal will go through a CCF to extract harmonics [27]. Finally, the grid impedance and inductive reactance values are calculated as follows:

$$\begin{aligned} u_{pcc}(h) &= u_\alpha^h + ju_\beta^h \\ i_{L2}(h) &= i_\alpha^h + ji_\beta^h \end{aligned} \tag{5.45}$$

$$\begin{aligned} R_g(h) &= \mathrm{Re}\left(\frac{u_{pcc}(h)}{i_{L2}(h)}\right) = \frac{u_\alpha^h i_\alpha^h + u_\beta^h i_\beta^h}{i_\alpha^h i_\alpha^h + i_\beta^h i_\beta^h} \\ L_g(h) &= \mathrm{Im}\left(\frac{u_{pcc}(h)}{i_{L2}(h)}\right) = \frac{u_\beta^h i_\alpha^h - u_\alpha^h i_\beta^h}{i_\alpha^h i_\alpha^h + i_\beta^h i_\beta^h}\frac{1}{\omega_h}. \end{aligned} \tag{5.46}$$

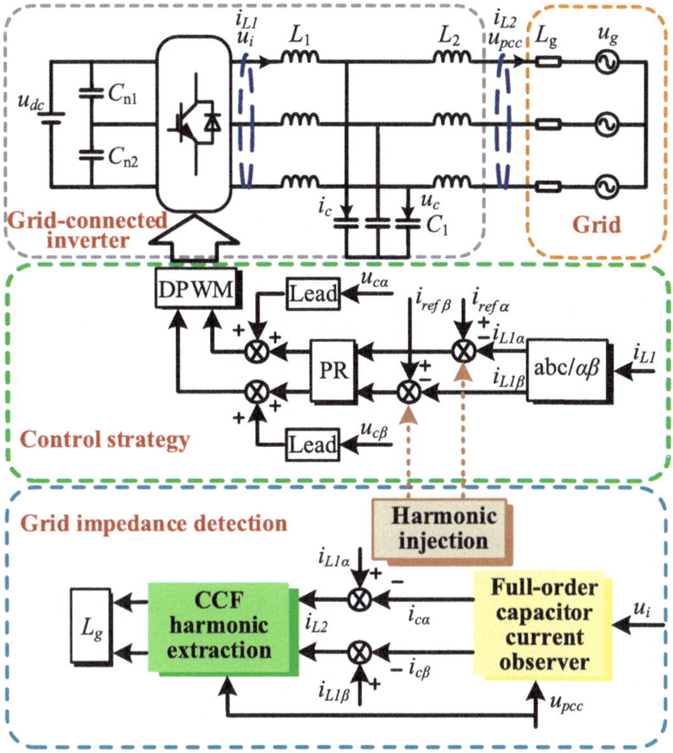

Fig. 5.23 Overall control block diagram of three-phase grid-connected inverter system

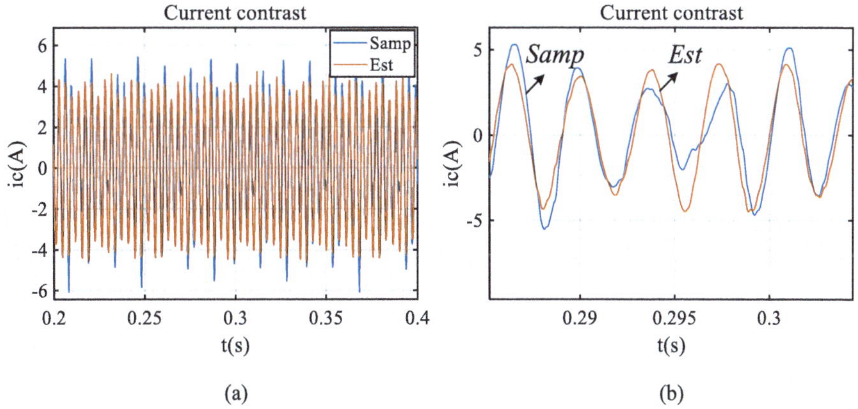

Fig. 5.24 Comparison results between the sampled capacitor current and the observed capacitor current after non-characteristic harmonic extraction when $L_g = 10\text{mH}$. **a** The phase-a sampled capacitor current. **b** The phase-a estimated capacitor current

5.4.4 Simulation Verification of Grid Impedance Detection Method Combining CCF and Current Observer

To verify the feasibility of the proposed method, the LCL three-phase grid-connected inverter model was built in MATLAB/Simulink. And the simulation control parameters are shown in Table 5.1. Through simulation analysis and experimental testing, it can be found that if the amplitude of injected harmonics is too large or the frequency of injected harmonics is near the fundamental frequency, the system stability will deteriorate. By contrast, if the amplitude of the injected harmonics is too small or the frequency of the injected harmonics is too high, the detection accuracy will be affected. The specific grid impedance detection parameters are shown in Table 5.2.

Comparison results between the sampled capacitor current and the observed capacitor current after non-characteristic harmonic extraction.

Figure 5.24 shows the comparison results of the phase-α sampled capacitor current and the observed capacitor current after non-characteristic harmonic extraction when the real grid impedance is $L_g = 10\text{mH}$. The right figure is the magnified image of the comparison results, where the blue line is the extraction result of the sampled value and the red line is the extraction result of the estimated value. According to the comparison results, it can be seen that after non-characteristic harmonic extraction, the observed capacitor current value is basically consistent with the sampled capacitor current value, thus verifying the correctness and effectiveness of the proposed full-order capacitor current observation method.

Due to the presence of high-frequency signals in the capacitor current, it is very difficult to accurately estimate it completely. After analysis and experiment, the slight error between the sampled capacitor current and the observed capacitor current has very little impact on the grid impedance detection results, thus verifying the feasibility of the proposed method.

Comparison results of grid impedance detection with and without considering capacitor current.

Figure 5.25 shows the comparison results of grid impedance detection with and without considering capacitor current when L_g is different (the left figure is the case of considering capacitor current, and the right figure is not considering capacitor current). Through the simulation results shown in Fig. 5.25, it can be seen that: (1)

Table 5.2 Simulation parameters of grid impedance detection

Parameter	Value	Parameter	Value
Injected harmonic amplitude	3 A	Injected harmonic frequency	275 Hz
Fundamental angular frequency ω_0	314 rad/s	Fundamental cutoff frequency ω_c	221 rad/s
Non-characteristic harmonic angular frequency ω_h	1728 rad/s	Non-characteristic harmonic ω_{hc}	500 rad/s

When the real grid impedance $L_g = 2\text{mH}$, the grid impedance detection value considering capacitor current is 2.02mH, and the grid impedance detection value without considering capacitor current is 2.075mH; (2) When the real grid impedance $L_g = 10\text{mH}$, the grid impedance detection value considering capacitor current is 9.95mH, and the grid impedance detection value without considering capacitor current is 11.6mH; (3) When the real grid impedance $L_g = 20\text{mH}$, the grid impedance detection value considering capacitor current is 20.3mH, and the grid impedance detection value without considering capacitor current is 27.5mH.

The comparison of simulation results from Fig. 5.25 shows that: (1) The error of grid impedance detection results with considering capacitor currents is controlled within 1.5%, and the error of grid impedance detection results without considering capacitor currents is controlled within 37.5%; (2) As the grid impedance increases, the error of the corresponding measured value without considering capacitor currents increases. Therefore, in order to realize the accurate detection of grid impedance of LCL type three-phase grid-connected inverters, it is necessary to estimate the capacitor current of three-phase grid-connected inverters.

Dynamic detection results of the grid impedance

The left figure of Fig. 5.26 shows the dynamic simulation detection results when the actual grid impedance value suddenly increases from 5 to 10 mH at 0.4 s. The simulation results of grid impedance detection change from 5.02 mH to 9.9 mH, and the detection error is less than 1%. The right figure of Fig. 5.26 shows the dynamic experimental detection results when the actual grid impedance value suddenly decreases from 10 to 5mH at 0.4 s. The simulation results of grid impedance detection change from 9.9 mH to 5.02 mH, and the detection error is less than 1%. Based on the above experiment results, it can be seen that the proposed method can accurately detect the grid impedance when the grid impedance is suddenly changed, the detection error is less than 1% and the detection time is about 0.15 s. As can be seen from the grid impedance waveforms in Figs. 5.25 and 5.26, the grid impedance detection method combining CCF and the full-order capacitor current observer proposed in this chapter can quickly and accurately detect the grid impedance.

5.4.5 *Conclusion of Grid Impedance Detection Method Based on Complex Coefficient Filter*

In order to ensure the stable operation of grid-connected inverters under both strong and weak grid conditions, this section studies the grid impedance detection method of LCL three-phase grid-connected inverters controlled by inverter current feedback. Firstly, the active detection method of non-characteristic harmonic injection is analyzed. Secondly, the principle of CCF and the specific structure of extracting non-characteristic harmonic from signals are analyzed. Thirdly, the grid impedance detection method combining CCF and full-order capacitor current observer is analyzed.

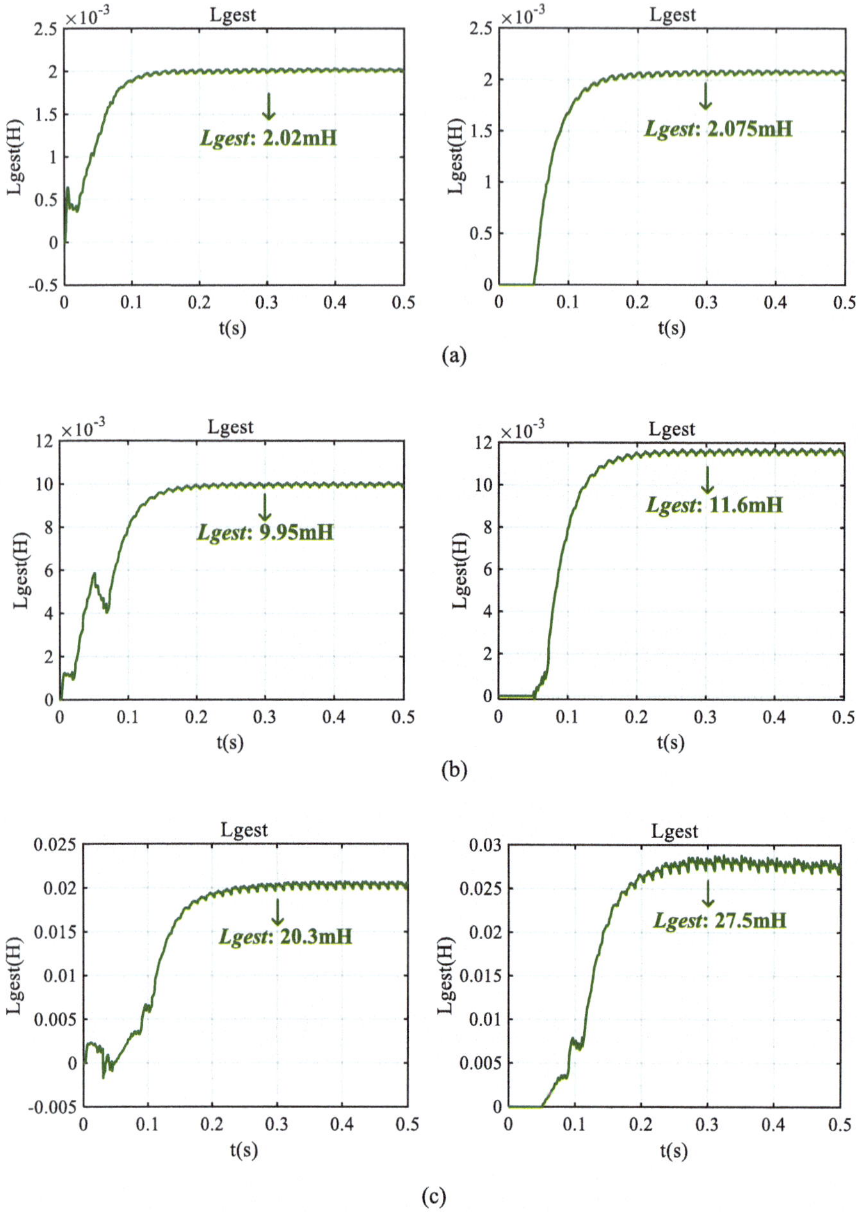

Fig. 5.25 Comparison results of grid impedance detection with and without considering capacitor current when L_g is different. **a** $L_g = 2$ mH. **b** $L_g = 10$ mH. **c** $L_g = 20$ mH

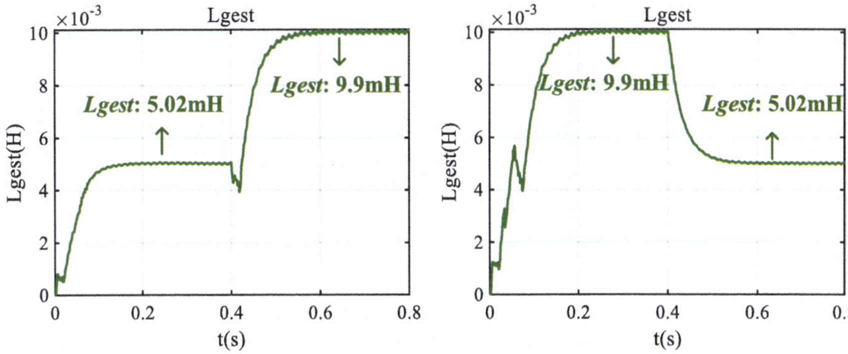

Fig. 5.26 Dynamic detection simulation results of the grid impedance

Finally, the correctness and effectiveness of the proposed grid impedance detection method combining CCF and full-order capacitor current observer are verified by MATLAB/Simulink simulation results.

5.5 Experiment on Three-Phase Grid-Connected Inverter and Impedance Detection with Complex Coefficient Filter Under Weak Grid

The experimental platform of the three-phase grid-connected inverter system established in this chapter is shown in Fig. 5.27 and the core control chip of the experimental platform is TMS320F28335 DSP, which is utilized to implement different control algorithms. Anymore, the experimental platform is composed of three-phase grid-connected inverters, TMS320F28335 DSP, inductor cabinet (which is used to adjust the impedance of the grid), DC power supply, and AC power supply. In the experiment, two 10 kW DC power supplies run in parallel to achieve the rated output power of 20 kW three-phase grid-connected inverters, AC source was used to simulate the ideal grid, the inductor cabinet was controlled to realize different SCR conditions, and the oscilloscope was used to observe the grid-connected voltage and current at the same time. The experimental parameters are consistent with the simulation parameters in Table 5.1.

5.5.1 Capacitor Voltage Lead Compensation Feedforward Comparison Experiment

According to the control strategy in Fig. 5.23, the experimental waveforms of phase-a common coupling point voltage and current and current THD under different L_g

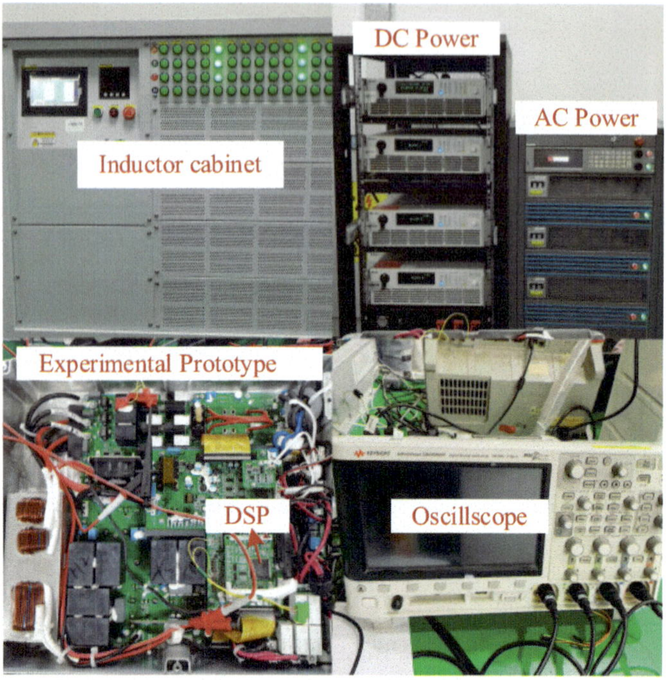

Fig. 5.27 Experimental platform of three-phase grid-connected inverter for weak grid

conditions are shown in Fig. 5.28 with the rated output power 20 kW (capacitor voltage Lead compensation is not added in the left figure while added in the right figure). It can be seen from the comparison experiment waveforms shown in Fig. 5.28:

(1) When the high impedance inductance of the grid is 0.1 mH and SCR = 23.12, the inverter can run stably with or without the capacitor voltage Lead compensation feedforward. The THD of the grid-connected current is 1.74% without the capacitor voltage Lead compensation feedforward and 1.76% with it. Therefore, adding or not adding capacitor voltage Lead compensation feedforward has no effect on the steady-state performance of grid-connected inverter with $L_g = 0.1$ mH.
(2) When the high impedance inductance of the grid is 10 mH and SCR = 2.312, the inverter can run stably when the capacitor voltage Lead compensation feedforward is added and the THD of the grid-connected current is 2.28%. The inverter will resonate and cannot run stably when the capacitor voltage Lead compensation feedforward is not added with $L_g = 10$ mH. The resonant frequency of a three-phase grid-connected current is around 2.273 kHz, which is approximately equal to the LCL resonant frequency of 2.43 kHz. This resonant frequency leads to the inverter system instability.

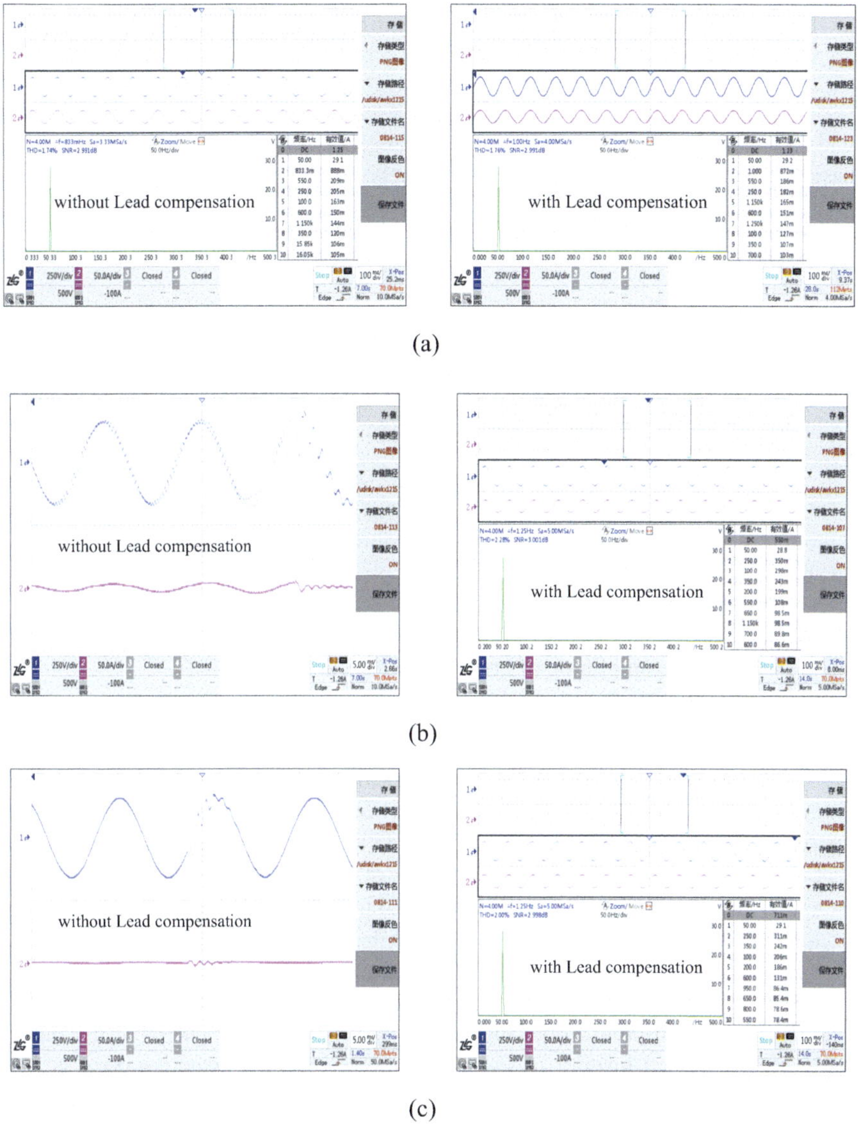

Fig. 5.28 Capacitor voltage Lead compensation feedforward comparison experiment with different L_g. **a** $L_g = 0.1$ mH. **b** $L_g = 10$ mH. **c** $L_g = 20$ mH

(3) When the high impedance inductance of the grid is 20 mH and SCR = 1.16, the inverter can run stably with the capacitor voltage Lead compensation feedforward and the THD of grid-connected current is 2%. When the capacitor voltage Lead compensation feedforward is not added, the voltage on the AC side of the grid-connected inverter will be too high at the moment of grid-connected

inverter. The LCL resonance phenomenon is more intense than $L_g = 10\text{mH}$, which is consistent with the theory.

The above three comparative experiments demonstrate the effectiveness of the Lead compensation feedforward algorithm to suppress LCL resonance, and the three-phase grid-connected inverter can operate stably under the extremely weak network condition with SCR = 1.16.

5.5.2 *Steady-State Experiment of Grid Impedance Detection*

To verify the effectiveness of the proposed grid impedance detection method combining CCF and capacitor current observer, experimental conditions are listed as follows:

Case A1: The output power of three-phase grid-connected inverter is 1 kW and the reference amplitude of grid-connected current is 2.14 A.
Case A2: The output power of three-phase grid-connected inverter is set to 5 kW and the reference amplitude of grid-connected current is 10.7 A.
Case A3: The output power of three-phase grid-connected inverter is set to 10 kW and the reference amplitude of grid-connected current is 21.4 A.

Figure 5.29 shows the experimental waveforms of phase-a common coupling point voltage u_{pcca} and grid-connected current i_{L2a} and phase-b grid-connected current i_{L2b} in case A1. According to the value recorded by the software oscilloscope, the grid impedance detection results are drawn in MATLAB. It can be seen from the experimental waveforms shown in Fig. 5.29: (1) When the real grid impedance $L_g = 2\text{mH}$, the impedance detection result is stable at 1.92mH and the detection error is 4%. (2) When the real grid impedance $L_g = 10$ mH, the impedance detection result is stable at 9.5 mH and the detection error is 5%. (3) When the real grid impedance $L_g = 20$ mH, the impedance detection result is stable at 19.7 mH, and the detection error is 1.5%. The above experimental results show that the proposed method can accurately detect different impedance values when the output power of the three-phase grid-connected inverter is 5%of rated power and the detection error is less than 5%.

Figure 5.30 shows the experimental waveforms of phase-a common coupling point voltage u_{pcca} and grid-connected current i_{L2a} and phase-b grid-connected current i_{L2b} in case A2. It can be seen from the experimental waveforms shown in Fig. 5.30: (1) When the real grid impedance $L_g = 2$ mH, the impedance detection result is stable at 1.985 mH and the detection error is 0.75%. (2) When the real grid impedance $L_g = 10$ mH, the impedance detection result is stable at 9.92 mH and the detection error is 0.8%. (3) When the real grid impedance $L_g = 20$ mH, the impedance detection result is stable at 20.7 mH, and the detection error is 3.5%. The above experimental results show that the proposed method can accurately detect different impedance

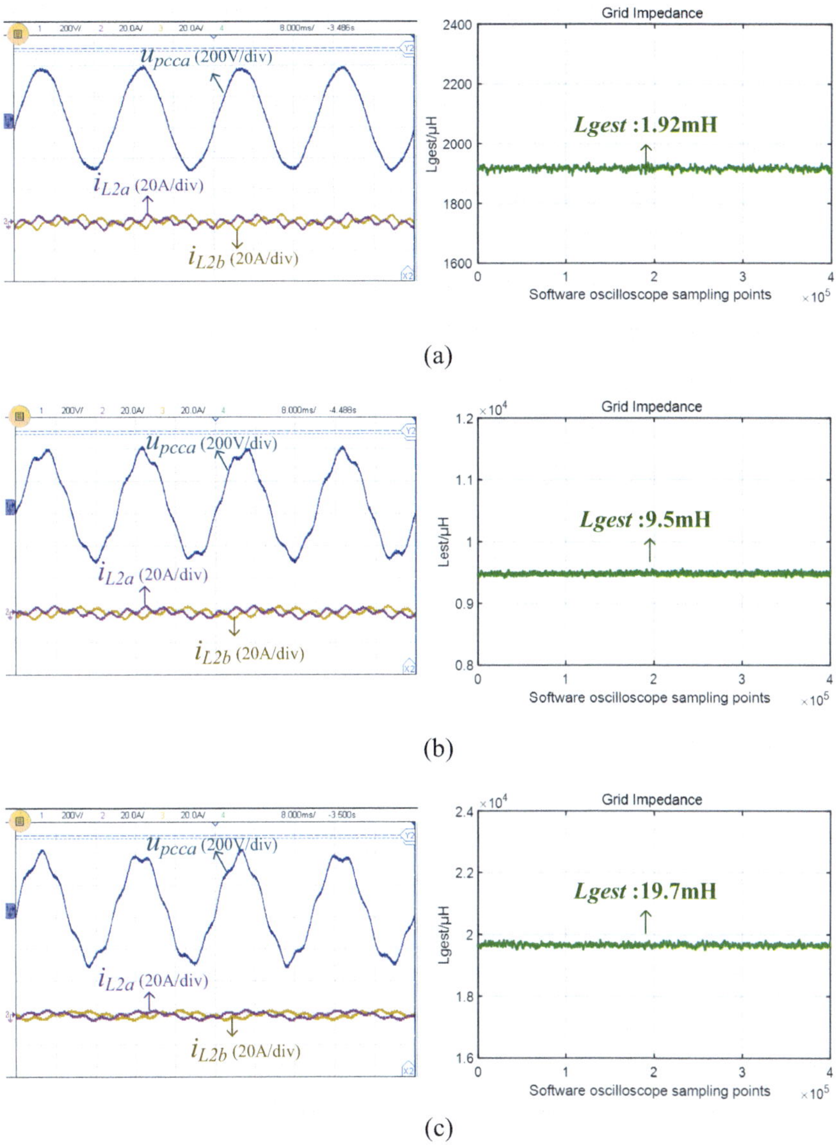

Fig. 5.29 Steady-state experimental results in case A1. **a** $L_g = 2$ mH. **b** $L_g = 10$ mH. **c** $L_g = 20$ mH

values when the output power of the three-phase grid-connected inverter is 25% of rated power and the detection error is less than 3.5%.

Figure 5.31 shows the experimental waveforms of phase-a common coupling point voltage u_{pcca} and grid-connected current i_{L2a} and phase-b grid-connected current i_{L2b}

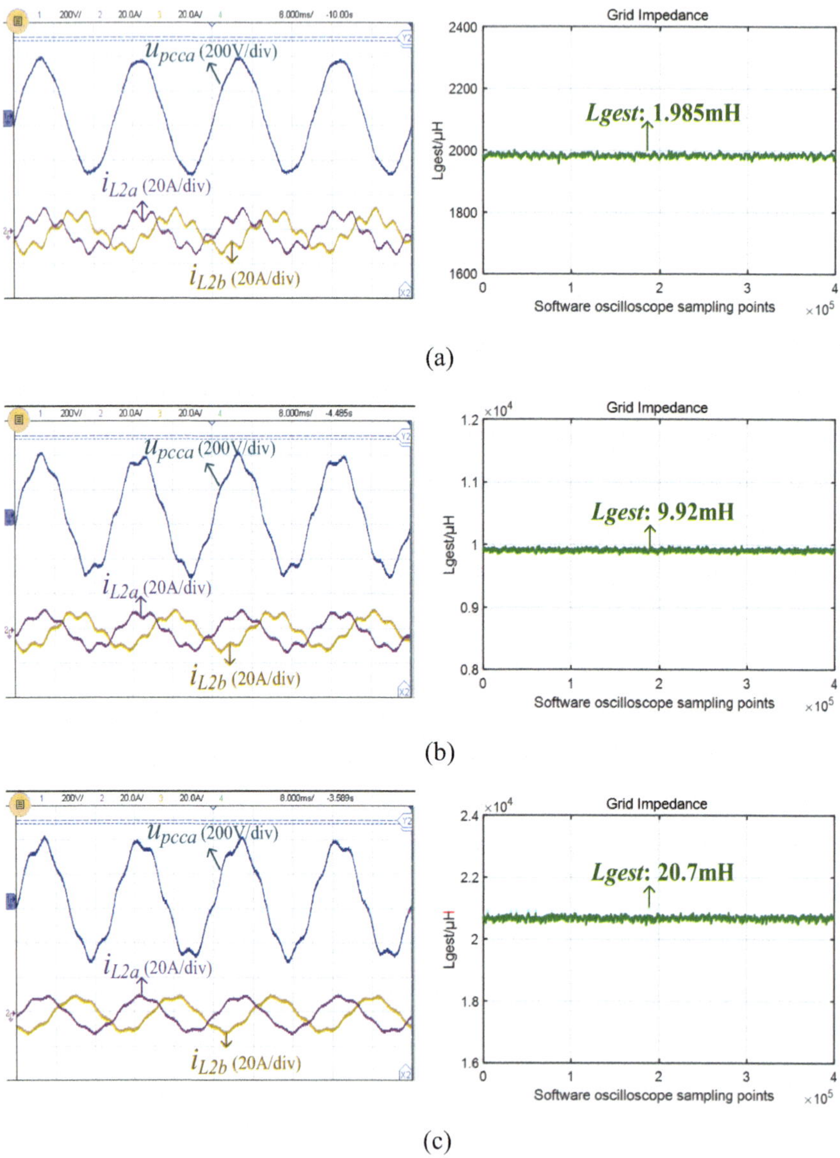

Fig. 5.30 Steady-state experimental results in case A2. **a** $L_g = 2$ mH. **b** $L_g = 10$ mH. **c** $L_g = 20$ mH

in case A3. It can be seen from the experimental waveforms shown in Fig. 5.31: (1) When the real grid impedance $L_g = 2$ mH, the impedance detection result is stable at 2.02 mH and the detection error is 1%. (2) When the real grid impedance $L_g = 10$ mH, the impedance detection result is stable at 10.18 mH and the detection error is 1.8%. (3) When the real grid impedance $L_g = 20$ mH, the impedance detection result is stable at 21 mH and the detection error is 5%. The above experimental results show that the proposed method can accurately detect different impedance values when the output power of the three-phase grid-connected inverter is 50% of rated power and the detection error is less than 5%.

5.5.3 Dynamic Experiment of Grid Impedance Detection

To verify the effectiveness of the proposed grid impedance detection method, dynamic experiments were conducted and the experimental conditions are listed as follows:

Case B1: The output power increases from 5 to 10 kW and the reference amplitude of grid-connected current increases from 10.7 A to 21.4 A;

Case B2: The output power decreases from 10 to 5 kW and the reference amplitude of grid-connected current decreases from 21.4 A to 10.7 A.

Figure 5.32 shows the experimental waveforms of phase-a common coupling point voltage u_{pcca} and grid-connected current i_{L2a} and phase-b grid-connected current i_{L2b} in case B1. It can be seen from the experimental waveforms shown in Fig. 5.32: (1) When the real grid impedance $L_g = 2$mH, the detection result changes from 1.99 mH to 2.02 mH, and the detection error is 1%. (2) When the real grid impedance $L_g = 10$ mH, the detection result changes from 9.92 mH to 10.1 mH, and the detection error is 1%. (3) When the real grid impedance $L_g = 20$ mH, the detection result changes from 20.5 mH to 21 mH, and the detection error is 5%. The above experimental results show that the proposed method can accurately detect different impedance values when the output power of three-phase inverter increases from 5 to 10 kW and the detection error is less than 5%.

Figure 5.33 shows the experimental waveforms of phase-a common coupling point voltage u_{pcca} and grid-connected current i_{L2a} and phase-b grid-connected current i_{L2b} in case B2. It can be seen from the experimental waveforms shown in Fig. 5.33: (1) When the real grid impedance $L_g = 2$mH, the detection result changes from 2.02 mH to 1.99 mH, and the detection error is 1%. (2) When the real grid impedance $L_g = 10$ mH, the detection result changes from 10.1 mH to 9.92 mH, and the detection error is 1%. (3) When the real grid impedance $L_g = 20$ mH, the detection result changes from 21 mH to 20.5 mH, and the detection error is 5%. The above experimental results show that the proposed method can accurately detect different impedance values when the output power of three-phase inverter decreases from 10 to 5 kW and the detection error is less than 5%.

It can be seen from the experimental results in Figs. 5.32 and 5.33 that when the output power of three-phase inverter increases or decreases by 50%, the proposed

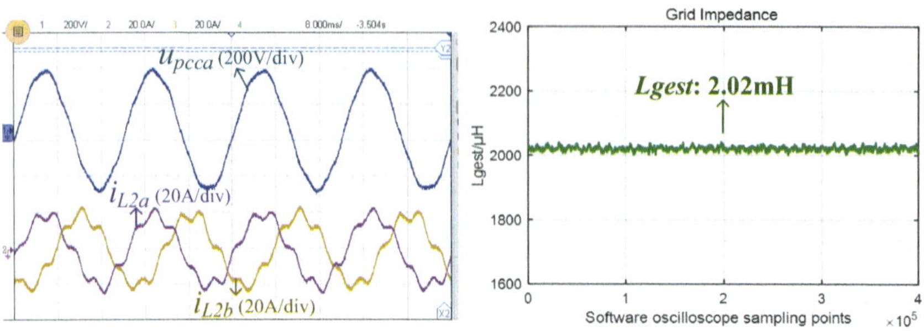

(a)

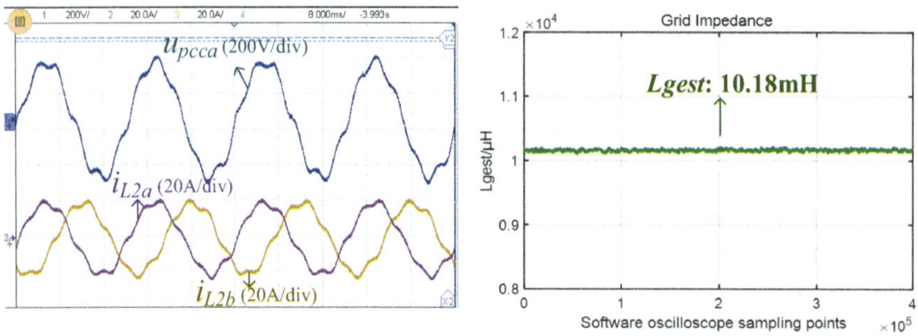

(b)

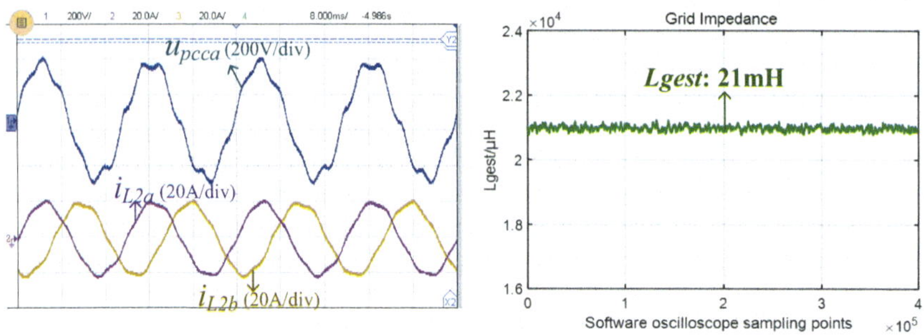

(c)

Fig. 5.31 Steady-state experimental results in case A3. **a** $L_g = 2$ mH. **b** $L_g = 10$ mH. **c** $L_g = 20$ mH

grid impedance detection method combining CCF and capacitor current observer in this chapter can detect grid impedance quickly and accurately.

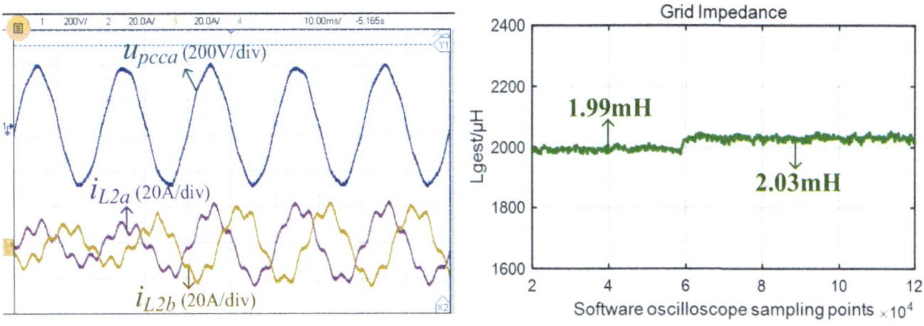

(a)

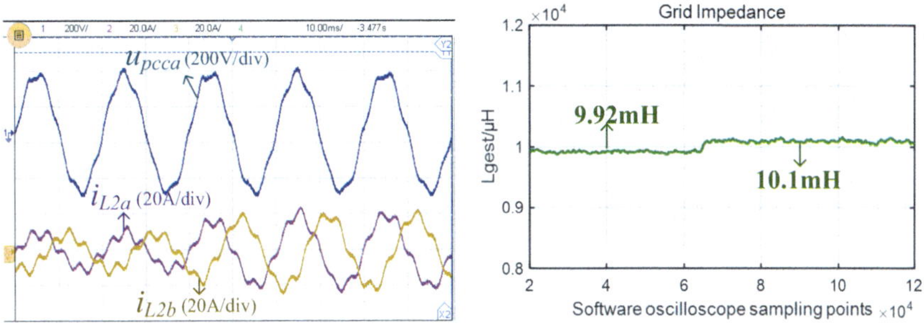

(b)

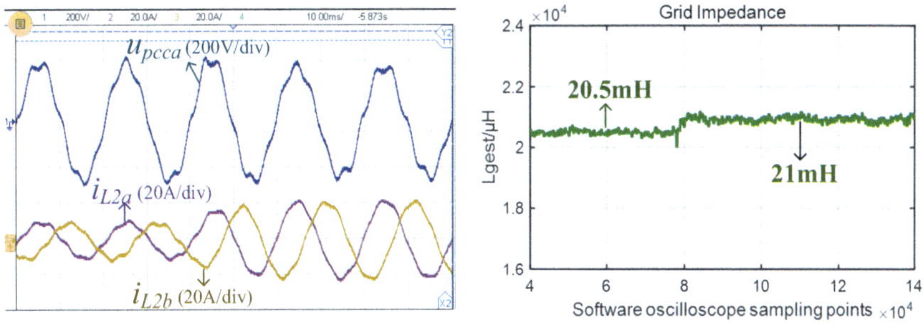

(c)

Fig. 5.32 Dynamic experimental results of grid impedance detection in case B1. **a** $L_g = 2$ mH. **b** $L_g = 10$ mH. **c** $L_g = 20$ mH

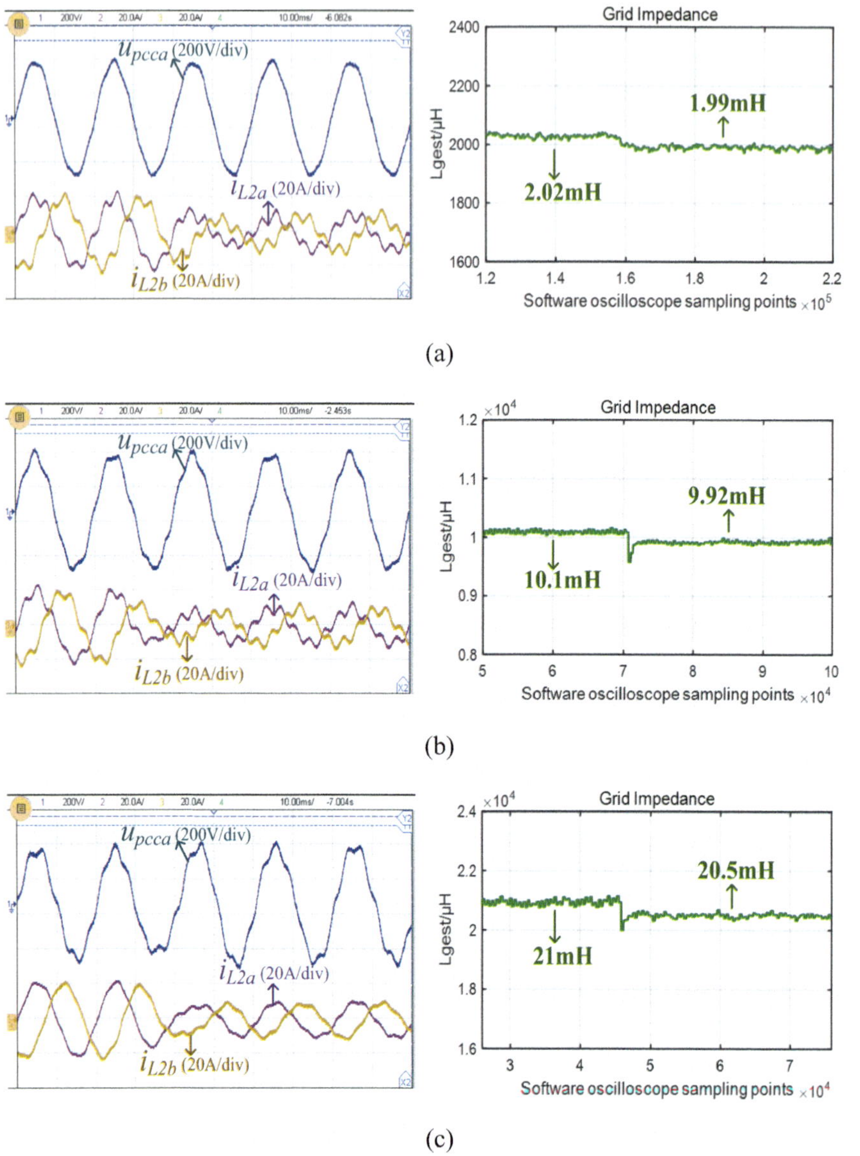

Fig. 5.33 Dynamic experimental results of grid impedance detection in case B2. **a** $L_g = 2$ mH. **b** $L_g = 10$ mH. **c** $L_g = 20$ mH

5.6 Summary of Impedance Detection Method for Three-Phase Grid-Connected Inverter Under Weak Grid

In this section, the experimental platform of a 20 kW three-phase grid-connected inverter is built to test the capacitor voltage Lead compensation method under a weak grid. Meanwhile, steady-state and dynamic experiments are carried out on the grid impedance detection method combined with CCF and capacitor current observer proposed in this chapter. The experimental results are consistent with the theoretical analysis results and the goal of the grid-connected inverter running stably and accurately detecting the grid impedance under the weak grid condition with SCR = 1.16 is realized.

5.7 Summary

Firstly, this chapter analyzes the research status of grid-connected control methods of inverters under weak power grids in community microgrids, then the capacitor voltage Lead compensation feedforward strategy is used to solve the problem of LCL resonance, and the grid-connected voltage control strategy is used to compensate reactive power to solve the problem of grid-connected voltage dropping. The effectiveness of the proposed method is verified by simulation and experiment. Besides, the grid impedance detection method based on CCF and full-order capacitive current observer is proposed. Simulation and experimental results show that the method can detect grid impedance quickly and accurately, and the detection error is less than 5%. The control method of the three-phase grid-connected inverter and the grid impedance detection method under a weak power grid provides a new idea for high-performance control of a distributed generation system.

References

1. X. Chen, Y. Zhang, S. Wang, J. Chen, C. Gong, Impedance-phased dynamic control method for grid-connected inverters in a weak grid. IEEE Trans. Power Electron. **32**(1), 274–283 (2017)
2. Z. Li, J. Hu, K.W. Chan, A new current limiting and overload protection scheme for distributed inverters in microgrids under grid faults. IEEE Trans. Ind. Appl. **57**(6), 6362–6374 (2021)
3. Z. Li, K.W. Chan, J. Hu, J.M. Guerrero, Adaptive droop control using adaptive virtual impedance for microgrids with variable PV outputs and load demands. IEEE Trans. Ind. Electron. **68**(10), 9630–9640 (2021)
4. J. Liu, X. Sun, B. Ren, W. Song, P. Wheeler, Strong adaptability control based on dual-division-summation current control for an LCL-Type grid-connected inverter. IEEE Trans. Power Electron. **37**(12), 14157–14172 (2022)
5. F. Cecati, R. Zhu, S. Pugliese, M. Liserre, X. Wang, State-Feedback reshaping control of voltage source converter. IEEE Trans. Power Electron. **37**(12), 14280–14293 (2022)

6. X. Wang, M.G. Taul, H. Wu, Y. Liao, F. Blaabjerg, L. Harnefors, Grid-synchronization stability of converter-based resources—an overview [J]. IEEE Open J. Indus. Appl. **1**, 115–134 (2020)
7. H. Wu et al., Small-signal modeling and parameters design for virtual synchronous generators [J]. IEEE Trans. Industr. Electron. **63**(7), 4292–4303 (2016)
8. Y. Yang et al., A novel continuous control set model predictive control for LC-filtered three-phase four-wire three-level voltage-source inverter [J]. IEEE Trans. Power Electron. **38**(4), 4572–4584 (2023)
9. J. Hu, Y. Shan, J.M. Guerrero, A. Ioinovici, K.W. Chan, J. Rodriguez, Model predictive control of microgrids—an overview. Renew. Sustain. Energy Rev. **136**(5), 1–12 (2021)
10. X. Fu, S. Li, Control of single-phase grid-connected converters with LCL filters using recurrent neural network and conventional control methods [J]. IEEE Trans. Power Electron. **31**(7), 5354–5364 (2015)
11. A.A. Rockhill, M. Liserre, R. Teodorescu, P. Rodriguez, Grid-filter design for a multimegawatt medium-voltage voltage-source inverter [J]. IEEE Trans. Industr. Electron. **58**(4), 1205–1217 (2011)
12. R. Peña-Alzola, M. Liserre, F. Blaabjerg, R. Sebastián, J. Dannehl, F.W. Fuchs, Analysis of the passive damping losses in LCL-filter-based grid converters [J]. IEEE Trans. Power Electron. **28**(6), 2642–2646 (2013)
13. J. Muhlethaler, M. Schweizer, R. Blattmann, J.W. Kolar, A. Ecklebe, Optimal design of LCL harmonic filters for three-phase PFC rectifiers [J]. IEEE Trans. Power Electron. **28**(7), 3114–3125 (2013)
14. S. Zhang, S. Jiang, X. Lu, B. Ge, F.Z. Peng, Resonance issues and damping techniques for grid-connected inverters with long transmission cable [J]. IEEE Trans. Power Electron. **29**(1), 110–120 (2014)
15. R. Peña-Alzola, M. Liserre, F. Blaabjerg, R. Sebastián, J. Dannehl, F.W. Fuchs, Systematic design of the lead-lag network method for active damping in LCL-filter based three phase converters. IEEE Trans. Industr. Inf. **10**(1), 43–52 (2014)
16. S. Cobreces, P. Rodriguez, D. Pizarro, F.J. Rodriguez, E.J. Bueno, Complex-space recursive least squares power system identification [C]. 2007 IEEE Power Electronics Specialists Conference, USA, pp. 2478–2484 (2007)
17. M. Ciobotaru, V. Agelidis, R. Teodorescu, Line impedance estimation using model based identification technique[C]. Proceedings of the 2011 14th European Conference on Power Electronics and Applications, UK, pp. 1–9 (2011)
18. N. Hoffmann, F.W. Fuchs, Minimal invasive equivalent grid impedance estimation in inductive-resistive power networks using extended Kalman filter [J]. IEEE Trans. Power Electron. **29**(2), 631–641 (2014)
19. Z. Xiangchen, Z. Guohui, Z. Jinbin, Impedance detection based on ripple analysis and current sharing control in DC microgrid [J]. IEEE Access **8**, 43554–43562 (2020)
20. D. Liang, H. Guo, T. Zheng, Real-time impedance estimation for power line communication [J]. IEEE Access **7**, 88107–88115 (2019)
21. D.K. Alves, R.L.A. Ribeiro, F.B. Costa, T.O.A. Rocha, Real-time wavelet-based grid impedance estimation method [J]. IEEE Trans. Industr. Electron. **66**(10), 8263–8265 (2019)
22. X. Guo, W. Wu, Z. Chen, Multiple-complex coefficient-filter-based phase-locked loop and synchronization technique for three-phase grid-interfaced converters in distributed utility networks [J]. IEEE Trans. Industr. Electron. **58**(4), 1194–1204 (2011)
23. Z. Wu, H. Zhang, K. Liu, W. Hua, G. Zhang, B. Wang, S. Ding, J. Hang, Dead-time compensation based on a modified multiple complex coefficient filter for permanent magnet synchronous machine drives [J]. IEEE Trans. Power Electron. **36**(11), 12979–12989 (2021)
24. J. Fang, H. Deng, S.M. Goetz, Grid impedance estimation through grid-forming power converters [J]. IEEE Trans. Power Electron. **36**(2), 2094–2104 (2021)
25. S. Neshvad, S. Chatzinotas, J. Sachau, Wideband identification of power network parameters using pseudo-random binary sequences on power inverters [J]. IEEE Trans. Smart Grid **6**(5), 2293–2301 (2015)

26. L. Asiminoaei, R. Teodorescu, F. Blaabjerg, U. Borup, Implementation and test of an online embedded grid impedance estimation technique for PV inverters [J]. IEEE Trans. Industr. Electron. **52**(4), 1136–1144 (2005)
27. K. Wang, Y. Yang, M. Fan, Y. Tang, H. Li, R. Chen, Hu., Jiefeng, W. Zeng, J. Rodriguez, Grid impedance detection based on complex coefficient filter and full-order capacitor current observer for three-phase grid-connected inverters [J]. IEEE Trans. Power Electron. **38**(2), 2408–2420 (2023)

Chapter 6
Load Probabilistic Forecasting Based on Neural Network

Haoxuan Chen and Yinliang Xu

6.1 Introduction

6.1.1 Intelligent Revolution of Power System

With the gradual development of the electricity production-to-usage structure in the community microgrid, the capabilities of grid operators need to be enhanced accordingly [1]. Figure 6.1 demonstrates the growing electricity consumption by household appliances and plug loads [2]. The diversification of grid-connected entities has brought difficulties in deploying electric energy and controlling the power system, represented by multiple loads. Specifically, the involvement of Electric Vehicles (EV), virtual power plants (VPP) as well as the complication of user behavior diversify the energy consumption patterns [3, 4]. Meanwhile, the rapid equipment of energy storage facilities and distributed generators leads to the diversity of electricity user identities, making them consumers and producers of electricity. Thus, the increasingly complex load with changeable characteristics brought difficulty to the system dispatching and planning.

In recent years, the integration of information technology and power grid operation has provided opportunities for the development of a smart grid. The penetration of uncertainty and randomness made load forecasting a core task in the intelligent development of the power grid [5]. Widely deployed monitoring sensors and massive historical data collected further advance the pace of realizing the load forecasting task. To meet the sustainable development goal of energy use, the scheduling of power systems can be optimized in advance through forecasting of load power

H. Chen · Y. Xu (✉)
Tsinghua-Berkeley Shenzhen Institute, Tsinghua Shenzhen International Graduate School, Tsinghua University, Shenzhen 518055, P. R. China
e-mail: xu.yinliang@sz.tsinghua.edu.cn

H. Chen
e-mail: chx23@mails.tsinghua.edu.cn

J. Hu (ed.), *Community Energy and Microgrids*, Green Energy and Technology,
https://doi.org/10.1007/978-981-97-6297-2_6

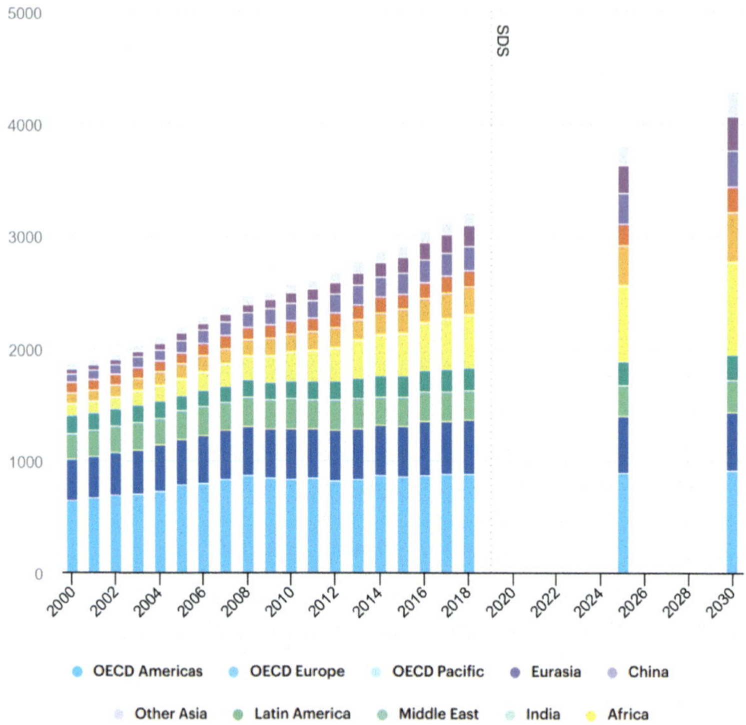

Fig. 6.1 Consumption by household appliances and plug loads by region

consumption level, which could guide demand response strategies and increase the efficiency of energy systems. Proper and effective load forecasting is the key to achieve unit commitment, generation scheduling, demand response [6], economic dispatch, energy trading [7], and other activities in community microgrids.

6.1.2 Development of Forecasting Technology

Existing research predicts the load by prediction models constructed from different aspects. Various time-series methods, regression methods, and machine learning methods have been proposed by researchers, which have been theoretically demonstrated and tested in real-world cases [8]. Load forecasting is a crucial topic regarding multiple detailed power system tasks, commonly classified from the following perspectives.

As shown in Fig. 6.2, load forecasting can be divided into various types according to sectors, time scales, models, and result forms. Load aggregators integrate decentralized demand response resources to participate in power system operations, which is a promising form of electricity load [9]. Additionally, different time scales of forecasting serve for diversified demand for smart grid operation, including designing, planning, dispatching, etc. Current mainstream forecasting models are constructed from perspectives including time-series methods, regression methods, and machine learning methods, where researchers modeled load from different angles to achieve the expected prediction goals [10]. With the development of forecasting technology and algorithms, researchers have successively proposed efficient forecasting models to solve the problem of power forecasting. The following is a review of load forecasting methods by model classification.

(1) Time-series methods

Time-series methods analyze historical load data to identify patterns and trends in electricity consumption over time. These methods are proposed based on the assumption that future load patterns will be similar to past patterns. The use of electric energy is closely related to human production activities, so the consumption of electricity

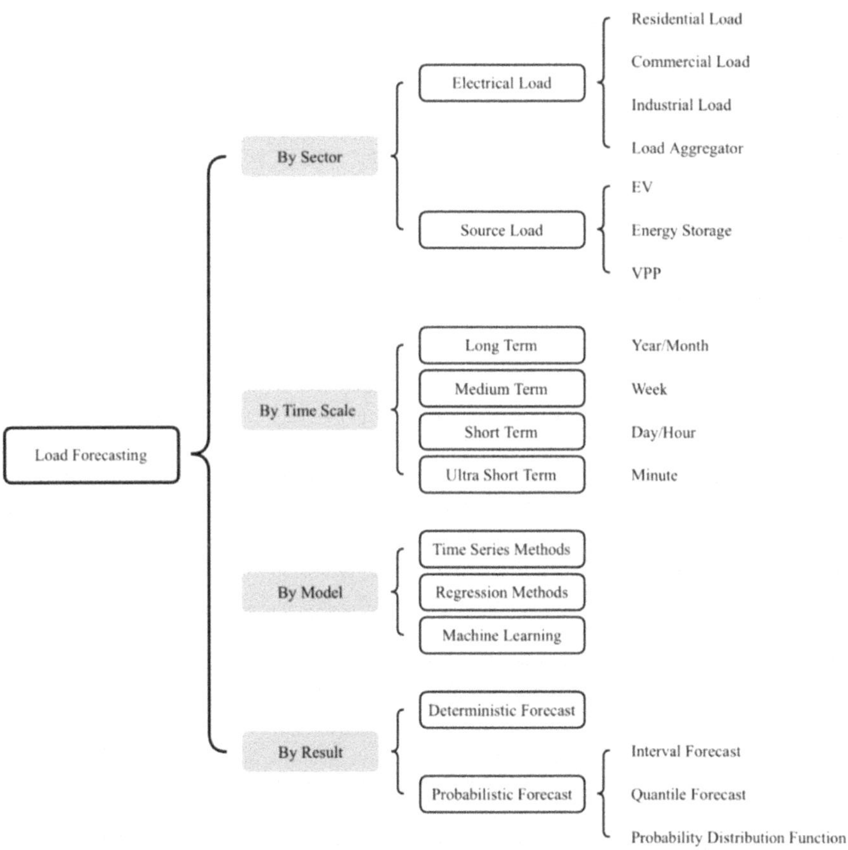

Fig. 6.2 Classification of load forecast

also fluctuates accordingly. Common load characteristics exhibit similar patterns in a daily period, in which curves present peaks and valleys accompanying people's production and life. The time-series model is a commonly used classic method, an autoregressive (AR) model using the linear combination of historical values and random disturbances [11]. The moving average (MA) model uses historical error weighting. Furthermore, the autoregressive moving average (ARMA) model integrates the first two methods and the autoregressive integrated moving average (ARIMA) model derived from non-stationary series enhanced adaptability of forecasting model in practical applications which is widely used in load prediction for its capability to predict sequence with trend or seasonality [12]. Through the construction of the above models, the quantification of trend, serial correlation, and randomness of time-series data can be realized. In addition, adaptive resampling and exponential smoothing were proposed to assign different weights to historical load values, and making detailed characterization for load patterns.

(2) Covariates regression methods

Covariates regression methods establish a statistical relationship between the load and other relevant factors that influence electricity consumption. The selected factors can include weather conditions, calendar variables, economic indicators, and population demographics [13]. Utilizing potential relationships between load and relevant covariates is a possible way to establish a connection that could improve forecasting performance. Specifically, numerical weather prediction is considered in a generalized Pareto distribution method for load forecasting [14]. Multiple linear regression is a widely used regression model for forecasting. Compared with forecasting models driven by single historical data, covariates enhance forecasting stability by establishing physical associations in electricity usage behavior [15].

(3) Machine learning methods

With the accelerated iteration of artificial intelligence technology, machine learning (ML), as the core of artificial intelligence, has given new capabilities to all walks of life. A series of machine learning-based methods have been proposed in energy forecasting. Due to their unique advantages in data mining and feature extraction, they provide nonlinear fitting capabilities superior to time-series and covariates regression methods for forecasting. Artificial neural network, support vector machine [16], extreme learning machine [17], backpropagation, and other shallow machine learning models were applied to electricity load forecasting. However, excessive parameters may increase the calculation time for the meta-heuristic optimization algorithm, and the mapping ability of the shallow ML framework limits its further development and application. As the main branch of ML, deep learning (DL) models are widely used in load forecasting [7]. Convolutional neural networks (CNN) can quickly extract key feature information in complex systems by using convolution kernels [18]. Recurrent neural network (RNN) adapts to the demands of processing the logical relationship of time-series data with a feedback structure and realizes feature extraction with a uniquely designed hidden layer structure [11]. Long short-term memory (LSTM) neural network was proposed to solve the long-term dependence problem, and the acquisition and evolution of its cell state enhanced the ability to deal with long-term sequence problems [19]. However, the outputs of classic ML models are sequences with point values, which cannot meet the demand from system operation in specific scenes.

6.1.3 Probabilistic Forecasting for Load

According to different forecasting results, load forecasting can be further divided into deterministic and probabilistic forecasting. The result of deterministic forecasting is a definite value of power, while probabilistic forecasting describes future electricity consumption in detail. The output of probabilistic forecasting includes the upper and lower bounds prediction information of power, the power consumption of different quantile points, or the entire probability density information [20].

Due to the complex structure of the user side of the power system and the large number of users, the load changes with significant randomness, intermittency, and volatility. Load probabilistic forecasting provides crucial information about the likelihood of high or low load conditions, helping grid operators prepare for potential fluctuations and take appropriate measures to ensure a stable grid [13]. Additionally, probabilistic forecasting allows utilities to communicate with consumers and encourage them to adjust their electricity consumption during high-load periods, helping to alleviate peak demand and reduce stress on the grid [21]. In deregulated electricity markets, load probabilistic forecasting could direct power trading decisions. Market participants, such as generators and retailers, use probabilistic information to manage their positions effectively and optimize their trading strategies based on forecasted demand. In summary, power load probabilistic forecasting provides critical insights and guidance for power system operators to make informed decisions, optimize resource utilization, enhance grid stability, and improve overall operational efficiency, emphasizing the importance of proposing corresponding forecast technology.

6.1.4 Highlights

Although load forecasting has made great progress in these years, there are still many technical difficulties to be overcome realizing the practical application of power grid load probabilistic forecasting in current research. Traditional point forecasting approaches provide a single prediction value, but they fail to capture the uncertainty inherent in power load data. To address this, a suitable probabilistic estimation method is required. Besides, the selection of relevant features from a vast pool of potential variables, such as weather conditions, historical load patterns, and economic factors significantly impacts the forecasting accuracy. Identifying the most influential features while minimizing noise and irrelevant information becomes crucial for building accurate and efficient forecasting models. Additionally, overcoming the strong data volatility problem requires the development of robust forecasting models capable of adapting to sudden fluctuations and providing reliable predictions under varying conditions.

In terms of the existing problems above, solutions are proposed in this chapter for load probabilistic forecast (LPF). The main highlights of this chapter are illustrated as follows:

(1) A load characteristic extraction approach is proposed for identifying potential features which is informative for forecasting. The series decomposition method is integrated for extracting deeper patterns of load.
(2) The load probabilistic forecasting is conducted by a novel hybrid network effectively generating future load information. The attention mechanism is applied to focus on influential tendency enhancing the forecasting stability while kernel density estimation is applied to generate probabilistic information.
(3) Considering strong volatility degrades model performance, the loss function is improved to tackle the deterioration problem. In comparison with existing methods, the accuracy increment of the proposed framework reaches 21.52–66.12%, embodying great potential in the system application.

6.2 Load Characteristics Extraction

Electricity consumption is influenced by various factors that could be quantified. Understanding the factors that influence electricity load is crucial for utilities and grid operators to develop accurate load forecasting models, plan resources effectively, and ensure stable and reliable power supply, which is demonstrated in Fig. 6.3. In this section, a load characteristic extraction (LCE) approach is proposed to obtain preliminary cognition of series characteristics and to obtain basic prior knowledge including series period, development trend, related variables, etc.

6.2.1 Formulation of the LCE

In the process of implementation, the LCE approach includes the decomposition method of strong volatility sequence based on Variational Mode Decomposition (VMD), feature selection method based on correlation, and time feature encoding method which extracts the feature of load from comprehensive aspects.

(1) System load model

Considering a distribution network with multiple buses which generates electricity consumption in a power system. The load aggregators collect and aggregate the electricity demand from multiple consumers to create a larger, more flexible load entity which allows smaller individual loads to be combined and managed as a single,

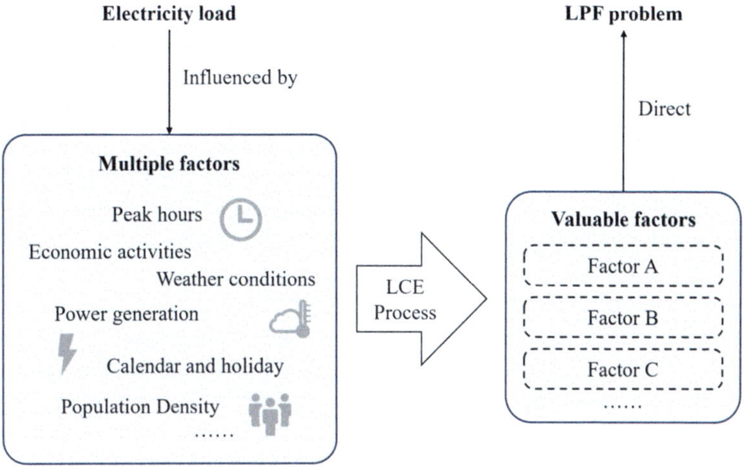

Fig. 6.3 Schematic diagram of LCE approach

larger load. In this study, the load aggregating process is not the research focus, thus the detailed process of aggregation is not discussed. Thus, the load of the regional power system can be generally defined as follows:

$$P^{\mathrm{Sys}}(t) = \sum_{i \in A} P_i^{\mathrm{Agg}}(t) \tag{6.1}$$

where A is the load aggregators set in the regional power system; $P^{\mathrm{Sys}}(t)$ is the total power consumption at time t; $P_i^{\mathrm{Agg}}(t)$ represents the power consumption for load aggregator i at time t.

Each load aggregator consists of numerous loads and the individual loads are combined as a superior load. The power for each load aggregator is modeled as follows:

$$P_i^{\mathrm{Agg}}(t) = \sum_{ir \in Lr} P_{ir}^{\mathrm{Res}}(t) + \sum_{ic \in Lc} P_{ic}^{\mathrm{Com}}(t) + \sum_{ii \in Li} P_{ii}^{\mathrm{Ind}}(t) + \sum_{io \in Lo} P_{io}^{\mathrm{Oth}}(t) \tag{6.2}$$

where Lr, Lc, Li, Lo represent the residential load, the commercial load, the industrial load, and the other load, respectively; $P_{ir}^{\mathrm{Res}}(t)$, $P_{ic}^{\mathrm{Com}}(t)$, $P_{ii}^{\mathrm{Ind}}(t)$, $P_{io}^{\mathrm{Oth}}(t)$ are the corresponding power consumption at time t.

The forecasting of the load is defined as a time-series problem. Thus, the power consumption and other variables are all time distributed:

$$\begin{cases} P = \{P(1), P(2), \ldots, P(t)\} \\ V = \{V(1), V(2), \ldots, V(t)\} \end{cases} \tag{6.3}$$

where $\boldsymbol{P}$ represents the load power consumption series; $\boldsymbol{V}$ represents the series of other variables related to load.

(2) Decomposition method based on VMD

The power system is a typical modern nonlinear complex system. There are often a variety of nonlinear effects, such as chaos, bifurcation, etc. These nonlinear effects will cause problems such as Gaussian noise and excessive spectral radius in the signal spectrum, making it difficult to process the power series. Since the raw load power series is a random sequence full of high-frequency noise and lacking regularity, it can be considered to decompose into several parts that are easier to analyze.

Given the fluctuant and less regular nature of the original power series, an efficient approach is to decompose it into a few easily analyzed components. Variational mode decomposition (VMD) serves as an effective algorithm for adaptively handling complex non-periodic signals [22]. It can efficiently capture various components, including oscillations, trends, and noise, present in the original series, gaining valuable insights into the underlying components of the original load series, which facilitates better analysis, prediction, and decision-making in power systems and electricity demand.

The decomposition of the original series involves several steps. Firstly, the Hilbert transformation is applied to each mode to obtain the unilateral frequency spectrum. Then, each frequency spectrum of the modal function is modulated to the corresponding baseband using exponential tuning. After that, Gaussian smoothing is performed to describe the bandwidth of each modal function. For a new energy power variation problem, the following formula can be obtained:

$$\left[\delta(t) + \frac{j}{\pi t}\right] * u_k(t) \tag{6.4}$$

To obtain the decomposition result of the power raw sequence, the Hilbert transform is first applied to each mode to obtain the one-sided spectrum. Then, each spectrum of the modal function is modulated to the corresponding baseband by exponential tuning.

$$\left\{\left[\delta(t) + \frac{j}{\pi t}\right] * u_k(t)\right\} e^{-j\omega_k t} \tag{6.5}$$

Gaussian smoothing was then performed to obtain bandwidth descriptions for each modal function. Supposed the power series is decomposed into K modes, the corresponding constrained variational problem is as follows:

$$\min_{\{u_k\},\{\omega_k\}} \left\{ \sum_{k=1}^{K} \left\| \partial_t \left\{\left[\delta(t) + \frac{j}{\pi t}\right] * u_k(t)\right\} e^{-j\omega_k t} \right\|_2^2 \right\} \tag{6.6}$$

$$s.t. \sum_{k=1}^{K} u_k = P_i^{Agg}(t) \tag{6.7}$$

where u_k is the kth mode; ω_k implicates the kth mode's center frequency. Symbol $\| \|\|_2^2$ represents the squared L2 norm; δ is the Dirac distribution and $*$ is the mark of convolution operations.

To solve this problem, the unconstrained problem variant is formulated using a quadratic penalty factor and a Lagrangian multiplier. This results in an augmented Lagrangian expression as follows:

$$\begin{aligned} L(\{u_k\},\{\omega_k\},\lambda) = \alpha \sum_{k=1}^{K} \left\| \partial_t \left[\left(\delta(t) + \frac{j}{\pi t} \right) * u_k(t) \right] e^{-j\omega kt} \right\|_2^2 \\ + \left\| P_{i,j}^{\text{gen}}(t) - \sum_{k=1}^{K} u_k(t) \right\|_2^2 \\ + \left\langle \lambda(t), P_i(t) - \sum_{k=1}^{K} u_k(t) \right\rangle \end{aligned} \tag{6.8}$$

where λ is the Lagrange multiplier, α is a quadratic penalty factor to ensure the accuracy of the series reconstruction. After the completion of the variational mode decomposition, the complex original sequence of load power is adaptively decomposed into multiple eigenmode function components for subsequent analysis. The pseudo-code for the algorithm execution is shown below.

Algorithm 1 Variational mode decomposition

Data: Load power series P, modal initial center constraint strength α_0, noise tolerance τ, number of modes, total signal fitting error tolerance

Result: Intrinsic Mode Function (IMF) Components

Initialize $\{\hat{u}_k^1\}, \{\hat{\omega}_k^1\}, \hat{\lambda}^1, n \leftarrow 0$;

repeat

  $n \leftarrow n+1$;

  for $k = 1{:}K$ do

    update $\hat{u}_k^1$ for all $\omega > 0$:

    $\hat{u}_k^{n+1}(\omega) \leftarrow \frac{\hat{P}(\omega) - \sum_{i<k} \hat{u}_i^{n+1}(\omega) - \sum_{i<k} \hat{u}_i^{n}(\omega) + 0.5\hat{\lambda}^n(\omega)}{1+2\alpha(\omega-\omega_k^n)^2}$

    update ω_k:

    $\omega_k^{n+1} \leftarrow \frac{\int_0^\infty \omega |\hat{u}_k^{n+1}(\omega)|^2 d\omega}{\int_0^\infty |\hat{u}_k^{n+1}(\omega)|^2 d\omega}$

  end

  dual ascent for all $\omega \geq 0$:

    $\hat{\lambda}^{n+1}(\omega) \leftarrow \hat{\lambda}^n(\omega) + \tau(\hat{P}(\omega) - \sum_k \hat{u}_k^{n+1}(\omega))$

until convergence $\sum_k \|\hat{u}_k^{n+1} - \hat{u}_k^n\|_2^2 / \|\hat{u}_k^n\|_2^2 < \varepsilon$.

(3) Feature selection method based on correlation analysis

The power load sequence data set contains a variety of auxiliary characteristic variables related to the physical world. Different physical characteristics have different influences on the actual power consumption. Different characteristics need to be identified and selected to ensure the computational efficiency of the network and eliminate the interference of irrelevant variables.

To further understand each auxiliary variable, the comprehensive serial correlation analysis method is adopted in this study to carry out variable screening. Different from the classic single-index correlation analysis method, this section uses the Pearson correlation coefficient, Spearman correlation coefficient, and maximum information coefficient as characteristic indicators for comprehensive research and judgment of correlation.

The Pearson correlation coefficient, namely the linear correlation coefficient, is used to detect the degree of linear correlation between two continuous variables, and its calculation formula is as follows:

$$\rho_P = \frac{\mathrm{Cov}(X_1, X_2)}{\sqrt{\mathrm{Var}(X_1)Var(X_2)}} \tag{6.9}$$

where X_1, X_2 are sequences to be estimated; $\mathrm{Cov}(\cdot)$ represents the calculation of covariance; $\mathrm{Var}(\cdot)$ represents the calculation of variance.

The Spearman correlation coefficient focuses on the monotonic relationship of variables, does not consider the influence of specific values, which can tolerate outliers, and does not require variables to obey a normal distribution. Its calculation formula is as follows:

$$\rho_S = 1 - \frac{6\sum_i d_i^2}{n\left(n^2-1\right)} \tag{6.10}$$

Among them, n represents the number of data points, and d_i represents the difference of rank i of data points.

The maximum information coefficient (MIC) can be used for nonlinear correlation analysis and can capture more types of functional relationships [23]. When calculating it, the two-dimensional scatter diagram composed of two characteristic variables is first gridded, then the maximum mutual information value will be computed. The calculation formula is as follows:

$$I^*(D, X_1, X_2) = \max I(D|_G) \tag{6.11}$$

where $D \in R^2$ is the maximum information entropy of a limited data set; G represents the symbol of the grid.

Then, each normalized MIC is calculated as follows:

$$\mathrm{MIC}(D) = \max_{X_1 * X_2 < B(n)} \left\{ \frac{I^*(D, X_1, X_2)}{\log(\min\{X_1, X_2\})} \right\} \tag{6.12}$$

where $B(n)$ is the superior limit of grid size; the length of data is defined as n; $\{dx, dy\}$ is the series to be compared.

After the correlation analysis, the results are weighted by multiple coefficients, which serve as the reference for selecting auxiliary variables. This selection process further determines the input of the forecasting model. It not only identifies valuable variables but also avoids unnecessary model complexity by excluding unrelated ones.

(4) Time feature encoding method

Electricity load is closely related to time features as it exhibits significant variations and patterns based on the time of day, day of the week, and season. Electricity load typically follows a diurnal pattern, meaning it shows distinct daily cycles. Specifically, residential loads tend to peak in the early mornings and evenings when people wake up and return home from work while commercial and industrial loads often peak during normal working hours and decrease during non-business hours. Thus, the hour information is encoded as a time feature sequence.

$$X_{t1} = \{\text{hour}(1), \text{hour}(2), \ldots \text{hour}(t)\} \tag{6.13}$$

Electricity load varies between weekdays and weekends. On weekdays, commercial and industrial loads contribute significantly, while residential loads play a more prominent role during weekends. Thus, the aggregated load presents a different pattern between days. To distinguish the day pattern, the weekday information is also encoded as a time feature sequence.

$$X_{t2} = \{\text{weekday}(1), \text{weekday}(2), \ldots \text{weekday}(t)\} \tag{6.14}$$

6.2.2 The Proposed LCE Methodology

Consisting of the above-mentioned components, the LCE approach is generalized to further excavate the feature of load pattern in the process of data preparation. Figure 6.4 illustrates the relationships and input/output information among the methods used in LCE, providing a more intuitive representation of the process.

The process of the proposed LCE approach is generalized as follows.

(i) Use the VMD algorithm to decompose the original load series to obtain IMFs with potential features;
(ii) Auxiliary variables are selected by comprehensive correlation analysis;
(iii) Time feature is encoded to provide extra information by data augmentation;
(iv) The IMFs and time feature series are added to the model auxiliary variables for tendency reference;

Thus, the electricity load prior knowledge is generated by the LCE approach, and more excavated information is used for model input.

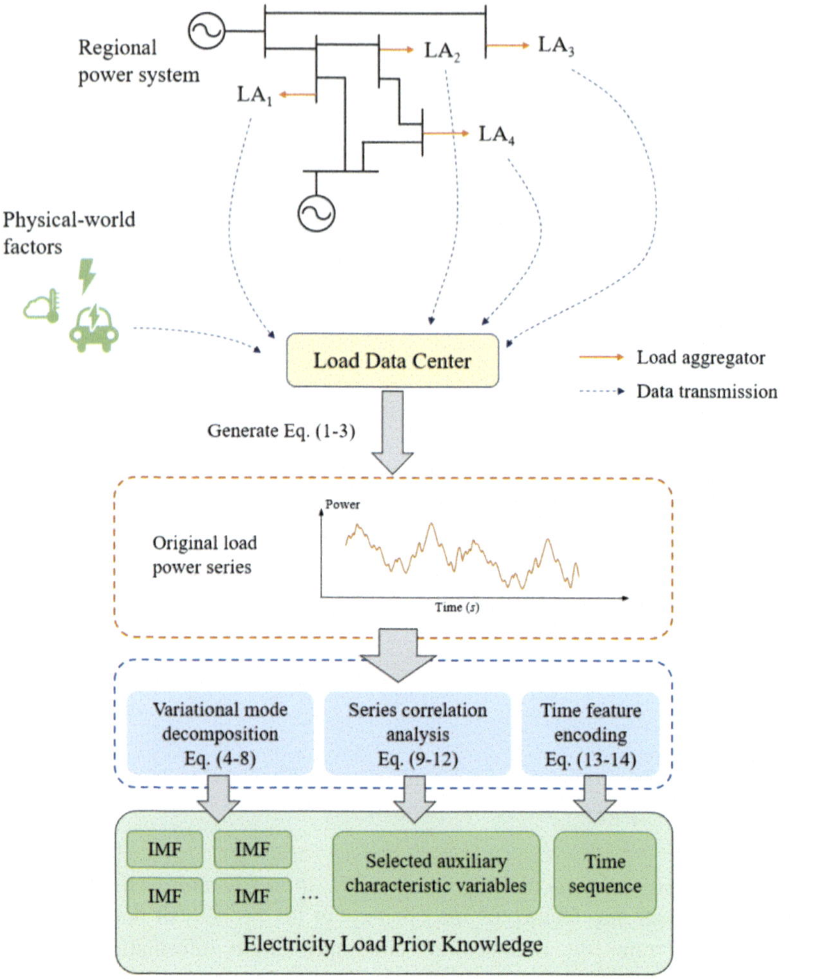

Fig. 6.4 The relationship between the used modules in LCE

6.3 Load Probabilistic Prediction Model

In order to meet the increasingly sophisticated requirements for power grid dispatching and power forecasting accuracy and reliability of results, researchers have successively proposed more complex and comprehensive forecasting models to meet the challenges faced at this stage. A typical trend is the large-scale application of artificial intelligence and neural network models, laying the foundation for the improvement of load forecasting effects. However, in the face of different forecasting demands, the forecasting model still needs targeted improvements to accomplish different tasks.

Typical improvement directions for the neural network include: (1) Improvement of the network structure: improve the efficiency and memory capacity of the network structure by improving the efficient network organization structure, and realize the interconnection of different functional modules through the connection relationship between the building modules; (2) Improvement of the loss function: change the convergence direction, process and speed of the network by designing a loss function with different characteristics, so as to improve the performance and results of the prediction model; (3) Combination of different modules: take advantage of the performance advantages of different modules, and improve the overall performance through combination methods.

This chapter will be organized as follows and will be divided into the following parts. First, the fundamental theory of neural network forecasting is introduced. Then, an attention-based sequence neural network is proposed for the load probabilistic forecasting method based on. Then, the model is improved for strong fluctuating loads, and an improved liberty loss function is proposed, which is used in load probabilistic forecasting.

6.3.1 Fundamentals of Neural Network Prediction

Neural networks have become the mainstream approach for time-series forecasting due to their ability to capture complex patterns and nonlinear relationships in the data. The advancements in flexibility and feature learning ensure the effectiveness of load forecasting applications, which outperform traditional methods in many aspects.

Recurrent Neural Network is a classic model in machine learning and the basis for many load forecasting technologies. The specific neural network structure takes generalized vector data as input and performs recursive calculations in sequence in the evolution direction, and all recurrent units (neurons) are connected in a chain. For each neuron of RNN, their parameters are always shared. For a historical time-series input, all input moments will be processed in the same way, generating the corresponding output. For the traditional fully connected network (FCN) structure, its neurons are independent of each other, where exists no direct connection, and they do not affect each other during the network iteration process. In the RNN, weight connections are also established between neurons between layers. In this type of neural network structure, the hidden layer is introduced, and the neurons in the hidden layer are connected through a hidden state. A typical RNN structure is shown in Fig. 6.5. The circles in the figure represent different types of values, including input, hidden layer, output, observation value, and loss. The left side of the figure shows the folded form of the RNN, and the right side shows the expanded form. The recurrent part of an RNN is embodied in the design of the hidden layers.

In the expanded form on the right, for the standard RNN structure, the neurons in the hidden layer are also connected by weights. That is, as the prediction sequence advances, the hidden layer will accumulate the information of the previous sequence and be influenced by it. In the figure, x represents the historical sequence of the input,

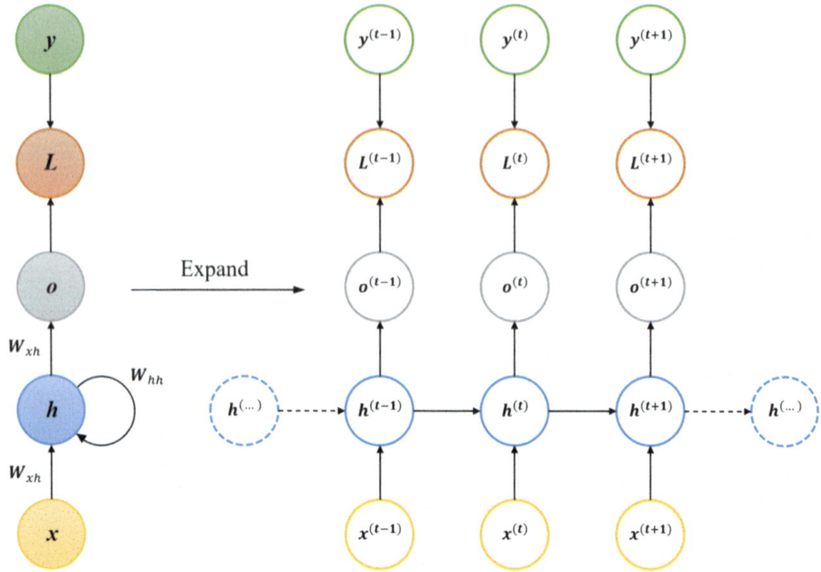

Fig. 6.5 The layer structure of RNN

h represents the hidden layer, o represents the output, y represents the sequence observation value, and L represents the loss function. With the training and learning of the historical load time series, the loss and network weights are constantly accumulating and changing, as detailed below.

(1) Forward propagation process

The input of RNN is fed into the network sequentially according to the sequence time sequence, and its forward propagation process is the evolution of the time. Among these, the state of the hidden layer in the RNN has a significant influence on the output of the network, and the connections between neurons in each hidden layer depend on the hidden state. In the specific propagation process in one layer of the neural network, the state $h_1^{(t)}$ for a time step t is obtained through the neurons in the hidden layer. At the next time step $t+1$, the neurons obtain the updated hidden state $h_1^{(t+1)}$ by receiving the output state of the hidden layer at the previous moment. Between layers, neurons are also connected through hidden states. Between layer L_1 and L_2, the corresponding hidden state $h_2^{(t)}$ receives the previous state $h_1^{(t)}$ as input and generate the updated hidden state in this layer. The formula of the forward propagation process is as follows:

$$h^{(t)} = \phi\left(W_{xh}x^{(t)} + W_{hh}h^{(t-1)} + b_h\right) \tag{6.15}$$

where $\phi(\cdot)$ is the symbol of the activation function and b_h is the bias. Generally speaking, tan$h$ can be selected as the activation function.

As shown in the above formula, for a certain layer, U, W, and b are the parameters that need to be updated during training, corresponding to the structure diagram of the RNN. The parameters of the recurrent units in the same layer are shared. The network output at time t can be expressed as

$$o^{(t)} = \sigma\left(W_{ho}h^{(t)} + b_o\right) \tag{6.16}$$

where $\sigma(\cdot)$ is the symbol of the activation function and b_o is the bias. Generally speaking, softmax can be selected as the activation function.

For the last layer of the network, the output prediction value is calculated as the following formula:

$$y^{(t)} = \sigma\left(W_{hy}h^{(t)} + b_y\right) \tag{6.17}$$

where b_y is the bias.

(2) Backpropagation process

Based on the forward propagation process of the above-mentioned network, the description of backpropagation is discussed. The core is to use the gradient descent method to obtain the appropriate model parameters W, b after repeated iterative calculations. Usually, the hidden state of the last time step of RNN is directly used as the output layer, the output activation function is *softmax* function, and the activation function of the hidden layer is *tanh* function. On this basis, the loss function is the sum of each moment:

$$L = \sum_{t=1}^{T} L^{(t)} \tag{6.18}$$

Therefore, for each neural network parameter, the update method is obtained according to the following gradient calculation. The relevant parameters of the output layer are as follows:

$$\frac{\partial L}{\partial W_{ho}}L = \sum_{t=1}^{T} \frac{\partial L}{\partial o^{(t)}}\frac{\partial o^{(t)}}{\partial W_{ho}} = \sum_{t=1}^{T}\left(\hat{y}^{(t)} - y^{(t)}\right)\left(h^{(t)}\right)^T \tag{6.19}$$

$$\frac{\partial L}{\partial b_o}L = \sum_{t=1}^{T} \frac{\partial L}{\partial o^{(t)}}\frac{\partial o^{(t)}}{\partial b_o} = \sum_{t=1}^{T} \hat{y}^{(t)} - y^{(t)} \tag{6.20}$$

During backpropagation, the parameters of the hidden layer, the gradient loss at a certain moment t of the sequence is jointly determined by the gradient loss corresponding to the output at the current moment and the gradient loss at $t + 1$ moment. As for the gradient loss of W at moment t, it requires the computation of backpropagation iterations. Thus, the gradient of the hidden state at time t is calculated as the following formula:

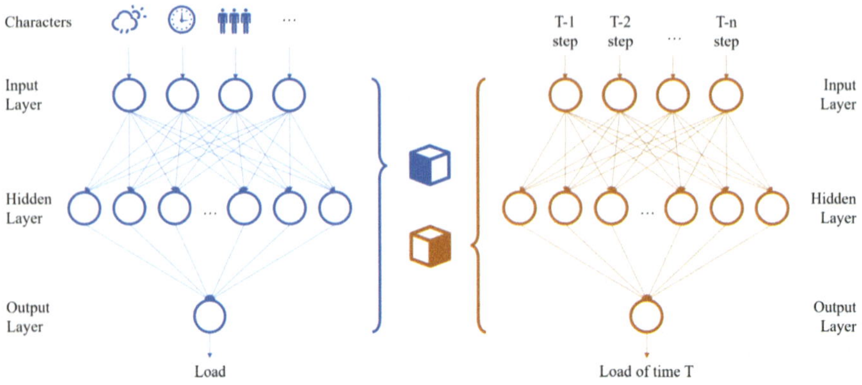

Fig. 6.6 Dimensions of load forecasting network

$$\delta^{(t)} = \frac{\partial L}{\partial h^{(t)}} = W_{ho}^T\left(\hat{y}^{(t)} - y^{(t)}\right) + W_{hh}^T\delta^{(t+1)}diag\left(1 - \left(h^{(t+1)}\right)^2\right) \tag{6.21}$$

The gradient of the relevant parameters of the hidden layer, the calculation formula is as follows:

$$\frac{\partial L}{\partial W_{hh}}L = \sum_{t=1}^{T}\frac{\partial L}{\partial h^{(t)}}\frac{\partial h^{(t)}}{\partial W_{hh}} = \sum_{t=1}^{T}\text{diag}\left(1 - \left(h^{(t)}\right)^2\right)\delta^{(t)}\left(h^{(t-1)}\right)^T \tag{6.22}$$

$$\frac{\partial L}{\partial b_h}L = \sum_{t=1}^{T}\frac{\partial L}{\partial h^{(t)}}\frac{\partial h^{(t)}}{\partial b_h} = \sum_{t=1}^{T}\text{diag}\left(1 - \left(h^{(t)}\right)^2\right)\delta^{(t)} \tag{6.23}$$

$$\frac{\partial L}{\partial W_{xh}}L = \sum_{t=1}^{T}\frac{\partial L}{\partial h^{(t)}}\frac{\partial h^{(t)}}{\partial W_{xh}} = \sum_{t=1}^{T}\text{diag}\left(1 - \left(h^{(t)}\right)^2\right)\delta^{(t)}\left(x^{(t)}\right)^T \tag{6.24}$$

(3) Multidimensional modeling

In LPF problem, due to the consideration of multidimensional load auxiliary data, the dimensionality of its neural network structure is expanded. Historical information is used as input for the forecasting model. In addition to the network time dimension, different auxiliary data types at the same time add another dimension. The illustration of multidimension is shown in Fig. 6.6.

6.3.2 Strong Volatility Improvement

Electricity load data typically exhibits high volatility due to various factors such as randomness of user behavior, weather changes, daily and seasonal patterns, economic activities, and special events. Alleviating the influence of volatility in electricity load data is essential for effective energy management, demand forecasting, and ensuring the stability of the power grid.

Before delving into time-series forecasting, few fundamental notations are established. In the case of a multivariate time series, we denote the input of the model as x, representing an $\theta \times \tau$ matrix. Each element, $x_t^{\theta} \in \mathbf{R}$ encompasses θ features over a time input length of τ. The goal of the forecasting framework is to generate a univariate prediction of load. Consequently, the actual future value can be expressed as

$$\begin{aligned} y &= (y_1, y_2, \ldots, y_l, \ldots, y_L) \\ &= \left(x_{\tau+1}^1, x_{\tau+2}^1, \ldots, x_{\tau+l}^1, \ldots, x_{\tau+L}^1\right) \end{aligned} \tag{6.25}$$

At the same time, the load prediction result of the framework is defined as $\hat{\boldsymbol{y}} = (\hat{y}_1, \hat{y}_2, \ldots, \hat{y}_l, \ldots, \hat{y}_L)$. The most frequently utilized loss function in traditional neural networks, Mean Squared Error (MSE), is computed as follows:

$$\text{MSE} = \tfrac{1}{L} \textstyle\sum_{i=1}^{L} \left(\hat{y}_i - y_i\right)^2 \tag{6.26}$$

While traditional loss functions like MSE and MAE are commonly effective in various models, they exhibit limitations when handling real-world data with strong volatility tendencies and multiple interfering factors. The underlying mapping relations in the input series may be obscured by these factors, making model training challenging. Being disturbed by irrelevant information, the model tends to deviate from the desired goal during iterations. Thus, the result commonly presents as a copy of the past value (CPV) since it achieves higher metric scores. A schematic diagram is shown in Fig. 6.7. However, the forecasting result, which is similar to the last observed value, is without reference value since it buries the effort of model construction and is insensitive to irregular change.

For a mathematical representation of this phenomenon, the following inequality is introduced as a criterion of CPV:

$$\sum_{i=1}^{L} \left(y_i - \hat{y}_i\right)^2 > \sum_{i=1}^{L} \left(y_{i-1} - \hat{y}_i\right)^2 \tag{6.27}$$

To alleviate CPV defect, the adjustment of the loss function is a direct solution direction since the loss function influences the iteration process significantly. Thus, the quantified amount of CPV should be punished in the loss function, that is

$$\sum_{i=1}^{L} \left[\left(y_i - \hat{y}_i\right)^2 - \left(y_{i-1} - \hat{y}_i\right)^2\right] \le 0 \tag{6.28}$$

Simultaneously, the MSE should be reduced to zero to minimize the error. Thus, if the variable in the above equation is replaced by the form of MSE, the following equation holds, supposing that the model is optimized:

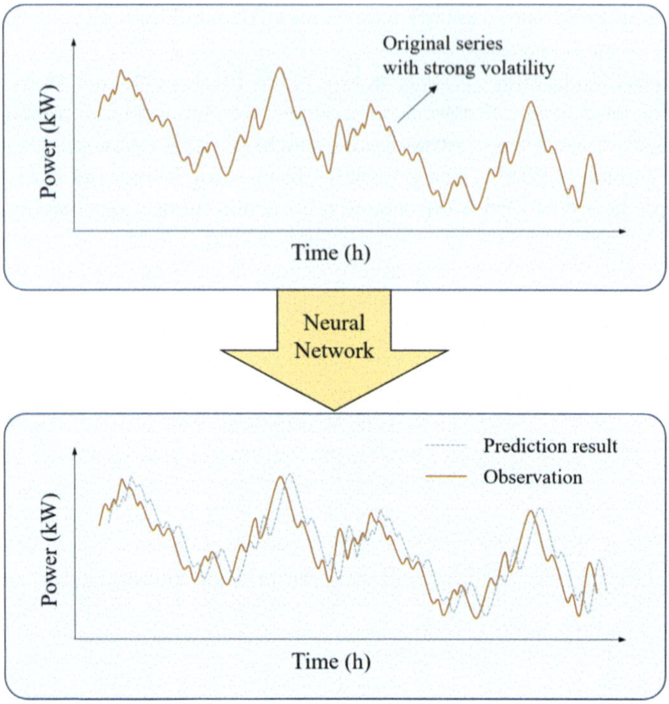

Fig. 6.7 The CPV phenomenon of neural network

$$0 - \sum_{i=1}^{L} \left(y_{i-1} - \widehat{y}_i\right)^2 = -\sum_{i=1}^{L} (y_{i-1} - y_i)^2 \tag{6.29}$$

To achieve optimality, the form to generate a lower bound of the loss function can be derived as follows:

$$-\sum_{i=1}^{L} (y_{i-1} - y_i)^2 \leq \sum_{i=1}^{L} \left(y_i - \widehat{y}_i\right)^2 - \sum_{i=1}^{L} \left(y_{i-1} - \widehat{y}_i\right)^2 \tag{6.30}$$

Combining (28) and (30), the ultimate form of the loss function avoiding the CPV phenomenon can be derived as a liberty loss function:

$$\mathcal{L}_{LB} = \sum_{i=1}^{L} \left(y_i - \widehat{y}_i\right)^2 + \lambda \sum_{i=1}^{L} \sum_{b=1}^{B} \left[(y_i - y_{i-b})\left(y_i - \widehat{y}_i\right)\right]^2 \tag{6.31}$$

where λ is an adjustable hyperparameter controlling the regularization term, which is the penalty intended to alleviate the CPV; B is the considered timesteps to prevent CPV during the period of B steps; $(y_i - y_{i-b})$, $\left(y_i - \widehat{y}_i\right)$ are two factors as cross-timestep penalty.

6.3.3 Probabilistic Forecast by Improved SANN

The main challenge in the load prediction framework is to establish a suitable model. Conventional neural networks often prioritize local trends and may overlook long-term features. To address this limitation, the sequence-based attention neural network (SANN) is proposed, leveraging its intricately designed structure to train and predict electricity load. The SANN incorporates three essential components: a sequence-based network, an attention mechanism, and a probabilistic prediction module, which collectively enable the consideration of the entire historical input while focusing on key characteristics.

(1) Sequence-based neural network

In practical load prediction, the multivariant and multistep prediction are simultaneously considered for power system operators. The historical electricity consumption data and physical world auxiliary data are all resources for forecasting. Therefore, both the input and output data are multidimensional vectors, which matches the encoder-decoder structure of the sequence-based model. The basic form of the sequence-based model is shown in Fig. 6.8. The model mainly consists of an encoder and a decoder. The encoder encodes the input sequence into a fixed-length vector, and the decoder decodes the vector into an output sequence. Each repeat unit for calculation is modeled using an LSTM cell.

In the encoder, the LSTM takes an input sequence and iteratively updates its hidden state vector, which captures the information in the input sequence. The final hidden state vector is defined as the encoder state, which is the input sequence for the decoder.

$$h_t^{\text{enc}} = \tanh\left(W\left[h_{t-1}, x_t\right] + b\right) \tag{6.32}$$

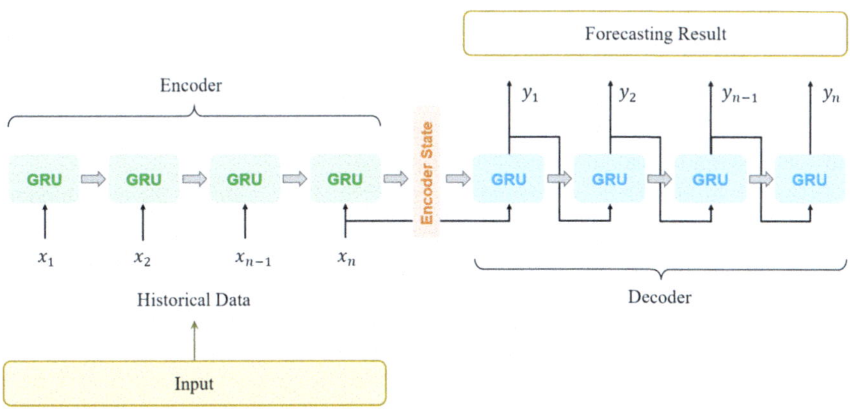

Fig. 6.8 The basic structure of sequence-based model

$$o_t^{\text{enc}} = \text{softmax}(Vh_t + d) \tag{6.33}$$

$$c = \tanh(Uh_T) \tag{6.34}$$

where x_t is defined as input sequence; h_t^{enc} is denoted as a hidden state vector of the encoder; o_t^{enc} is denoted as an output of the LSTM cell in the encoder; c denotes the encoder state; U, V, W is the weight matrix of each part; b, d is the normalization bias; the symbol $\tanh(\cdot)$ and $\text{softmax}(\cdot)$ represents the activation function.

In the decoder, each LSTM calculates the next output by taking the encoder state and the output of the previous time step. The process is repeated until the complete target sequence is generated.

$$h_t^{\text{dec}} = \tanh\left(W\left[h_{t-1}, y_{t-1}, c\right] + b\right) \tag{6.35}$$

$$o_t^{\text{dec}} = \text{softmax}(Vh_t + d) \tag{6.36}$$

where y_t represents the historical target sequence; h_t^{dec} is denoted as the hidden state vector of the encoder; o_t^{dec} is denoted as an output of the LSTM cell in the encoder.

(2) Attention Mechanism

The relationship between future load power consumption and historical data could be interpreted as a complex nonlinear mapping process, where different time steps in the past and different auxiliary variables make unbalanced contributions to the future output. Although the sequence-based model has overcome the long-term gradient vanishing problem, the decoder relies on the single invariable output vector which restricts the model from focusing on valuable information. Attention is a novel mechanism that is derived from the habit the human brain works which ensures the model focuses on relevant parts of the input sequence. It enables the model to process long sequences more accurately while improving the model's generalization ability and adaptability to complex tasks. Thus, the attention mechanism is designed in the above model.

The attention mechanism makes adjustments in the decoder method. Specifically, the attention layer is added to the existing network. The target of the attention layer is to iterate the similarity weight matrix based on the output of the encoder layer. The concrete mathematical formula is as follows:

Given the hidden state of the encoder as $(h_1^{\text{enc}}, h_2^{\text{enc}}, \ldots, h_T^{\text{enc}})$, for each hidden state of the decoder h_{t-1}^{dec}, the correlation between each input position j and the current output position is calculated by

$$e_{tj} = \tanh\left(\text{attn}\left(h_{t-1}^{\text{dec}}, h_j^{\text{enc}}\right)\right) \tag{6.37}$$

$$e_t = v^T(e_{t1}, e_{t2}, \ldots, e_{tT}) \tag{6.38}$$

where e_{tj} represents the correlation value, the vector form is denoted as e_t; attn(·) is defined as the correlation operator; v^T is the transpose of the weight vector.

The attention weight distribution is normalized by the calculation below:

$$\alpha_{tj} = \frac{\exp(e_{tj})}{\sum_{k=1}^{T} \exp(e_{tk})} \tag{6.39}$$

$$\alpha_t = \text{softmax}(\boldsymbol{e}_t) \tag{6.40}$$

where α_{tj} represents the attention weight and the vector form is denoted as α_t, exp(·) is the symbol of the exponent operator.

Then, the context vector is calculated from the attention weight and hidden state of the encoder. In the decoder, the output of the encoder is first defined as c_0, then c_t is continuously updated for each decoder step:

$$c_t = \sum_{j=1}^{T} \alpha_{tj} h_j^{\text{enc}} \tag{6.41}$$

Thus, the hidden state of the decoder is calculated and the output of the network is generated by the output gate.

$$h_t^{\text{dec}} = \tan h\left(W\left[h_{t-1}, y_{t-1}, c_t\right] + b\right) \tag{6.42}$$

$$y_t = g(y_{t-1}, s_t, c_t) \tag{6.43}$$

where $g(\cdot)$ obeys the rules of LSTM cell, which could calculate the output result.

The overall structure of the attention mechanism is demonstrated in Fig. 6.9. Improved by attention, the network model is designed to deal with long sequences with reinforced capability.

(3) Probabilistic Prediction Module

The expected probabilistic result is then generated by the estimation approach. Kernel Density Estimation (KDE) is used in this study which is a nonparametric estimation method. KDE works by placing a kernel at each data point and then summing up the contributions from all kernels to obtain an estimate of the underlying probability distribution. The width of the kernel, known as the bandwidth, controls the smoothness of the estimation, impacting the trade-off between capturing fine details in the data and obtaining a more general representation of the distribution. The formula is as follows:

$$\widehat{f}_h(x) = \frac{1}{nh} \sum_{i=1}^{n} K\left(\frac{x - x_i}{h}\right) \tag{6.44}$$

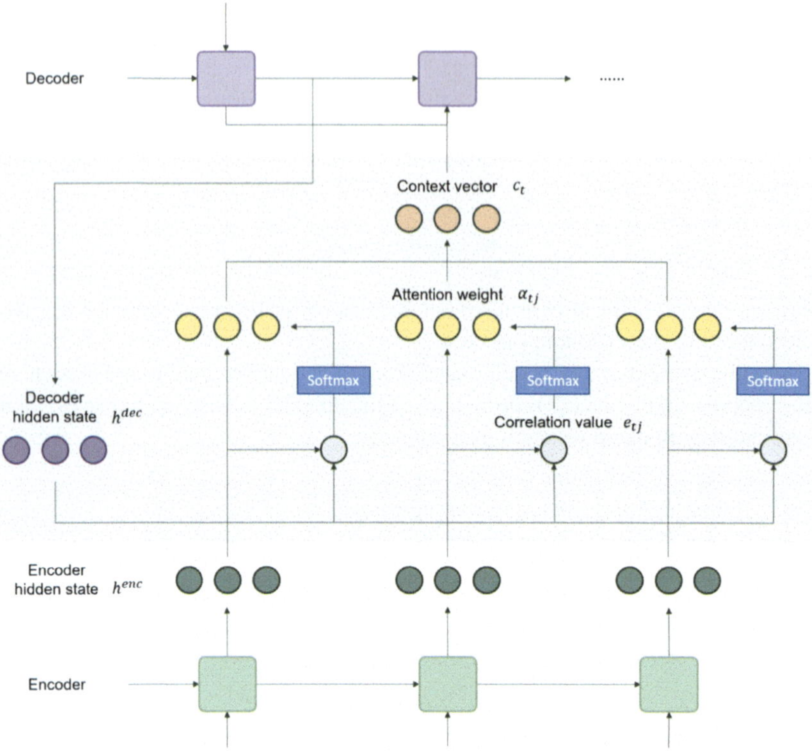

Fig. 6.9 The structure of the attention mechanism in sequence-based model

where n is the sample amount; h is the bandwidth, Silverman rule is used to determine the bandwidth; $K(x)$ is a kernel function, Gaussian kernel is used in this study.

(4) Framework of Load Probabilistic Prediction Approach

Given the modules above, the overall proposed framework for the probabilistic LPF problem is summarized as follows.

The load power sequence is obtained as a prediction target. The framework input is first processed by the LCE approach. Specifically, the load sequence is decomposed by VMD while the original physical world features are selected in advance to filter the irrelevant variables. The time features are also encoded which are used as covariates for model input.

Then, the SANN is trained by historical data when model parameters converge to the optimal as iterating. Meanwhile, the probabilistic forecast is estimated. Further, the prediction result is evaluated by comprehensive evaluation metrics to examine the effectiveness of the proposed framework. Parameter tuning is adopted until the performance is optimized. Eventually, the deterministic and probabilistic result is generated. The relation between the used modules in the load probabilistic forecasting framework is presented in Fig. 6.10.

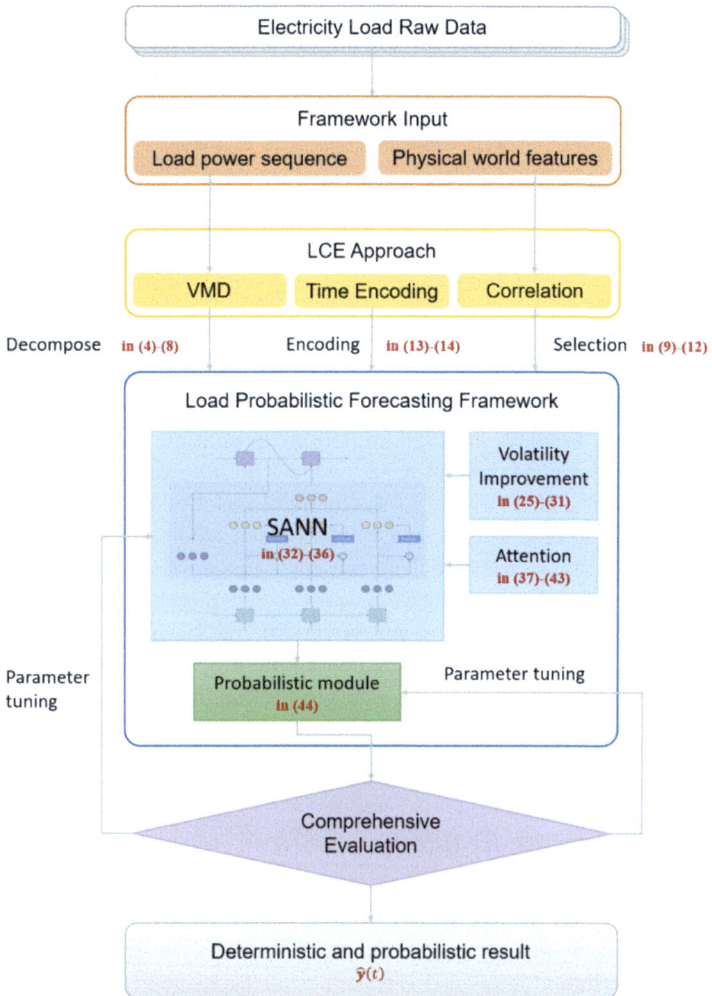

Fig. 6.10 The structure of load probabilistic forecasting framework

6.4 Case Study

6.4.1 Dataset Description and Environment Setup

In this chapter, an ensemble real-world recorded datasets of an aggregated electricity load in Panama with data sequences consisting of historical load and multiple physical world factors are used in the case study. The electricity load datasets with 15-min resolution covering four seasons from January 1, 2016, to December 31, 2016, are tested to detect annual atmospheric patterns. The dataset contains several features

Table 6.1 Hyper parameters of proposed model

Parameters	Value
Optimizer	Adam
Train mode	Teacher forcing
Loss function	Liberty loss
Learning rate	3e−4
Batch size	128
RNN size	64
Dense size	32
Layer num.	8

recorded by institutions and websites, including temperature, humidity, liquid precipitation, wind speed, holidays, and school periods in multiple areas. The historical sequence of 24 h is used as input to generate the 1 h future value. For the segmentation of the dataset, eighty percent of the entire electricity load dataset is used as a training set and the rest is used as a validation set. Additionally, the optimized hyperparameter settings of the model are presented in the following Table 6.1.

The load probabilistic forecasting model proposed in this study was implemented under Keras in a Python environment and MATLAB 2022b with Windows 11 Home Edition. All experiments were implemented on a desktop computer with 12th Gen Intel^{R} Core™ i9-12900H clocking at 2.50 GHz, NVIDIA GeForce RTX™ 2050 GPU, and 16.0 GB of RAM.

6.4.2 *Comprehensive Evaluation Metrics*

In electricity load power sequence prediction, the performance of the proposed model can be evaluated from various aspects below: the accuracy metrics for determined prediction, the prediction interval (PI) reliability metrics, the PI sharpness metrics, and the comprehensive probabilistic metrics. Definitions of these evaluation metrics are introduced as follows:

Accuracy metrics for determined prediction: The target of determined prediction is minimizing the deviation between observation and predicted value. The root mean square error (RMSE) and mean absolute percentage error (MAPE) are selected, which are defined as

$$\mathrm{RMSE} = \sqrt{\frac{1}{N}\sum_{i=1}^{N}\left(\hat{y}_i - y_i\right)^2} \tag{6.45}$$

$$\mathrm{MAPE} = \frac{1}{N}\sum_{i=1}^{N}\left|\frac{\hat{y}_i - y_i}{y_i}\right| \times 100\% \tag{6.46}$$

where N is the maximum length of time; $\hat{y}_i, y_i$ is the prediction result and actual value of the problem.

The standard deviation of error (SDE) is also introduced for determined evaluation, which is defined as follows:

$$\delta_i = y_i - \hat{y}_i \tag{6.47}$$

$$\text{SDE} = \sqrt{\frac{1}{N}\sum_{i=1}^{N}\left(\delta_i - \sum \frac{\delta_i}{N}\right)^2} \tag{6.48}$$

where δ_i represents the residual of time step i, the SDE is calculated by considering the overall residuals.

Notably, the CPV problem caused by strong volatility is evaluated by evaluation metrics LIB. A higher LIB indicates a more significant CPV phenomenon. The formula is defined as follows:

$$\text{LIB} = \frac{1}{N}\sum_{i=1}^{N}\left[\left(y_i - \hat{y}_i\right)^2 - \left(y_{i-1} - \hat{y}_i\right)^2\right] \tag{6.49}$$

PI reliability metrics: Since the probabilistic prediction is to estimate the uncertainty. The reasonable estimate of risk should be evaluated by metrics. The coverage probability of PI (PICP) is used to assess the percentage of points within the interval, which is defined as

$$\theta_i^{(\alpha)} = II\left(\hat{y}_i \in \left[\text{LB}_i^{(\alpha)}, \text{UB}_i^{(\alpha)}\right]\right) \tag{6.50}$$

$$\text{PICP}^{(\alpha)} = \frac{1}{N}\sum_{i=1}^{N}\theta_i^{(\alpha)} \times 100\% \tag{6.51}$$

where $\theta_i^{(\alpha)}$ is a bool number representing whether sample i is in the prediction interval with α confidence level; $\text{LB}_i^{(\alpha)}$, $\text{UB}_i^{(\alpha)}$ is the upper and lower bound of the interval with a confidence level of α; $\text{II}(\cdot)$ is the symbol representing the indicative function.

The average coverage error (ACE) is defined to measure the deviation between PI nominal confidence (PINC) and predicted PICP. The lower ACE indicates higher reliability, its formula is as follows:

$$\text{PINC}^{(\alpha)} = (1 - \alpha)\% \tag{6.52}$$

$$\text{ACE} = \text{PCIP}^{(\alpha)} - PIN\text{C}^{(\alpha)} \tag{6.53}$$

PI sharpness metrics: The sharpness of PI is commonly evaluated by PI normalized average width (PINAW) and PI mean width percentage (PIMWP), which is defined as

$$\text{PINAW}^{(\alpha)} = \frac{1}{N(y_{\max} - y_{\min})} \sum_{i=1}^{N} \text{UB}_i^{(\alpha)} - \text{LB}_i^{(\alpha)} \tag{6.54}$$

$$\text{PIMWP}^{(\alpha)} = \frac{1}{N} \sum_{i=1}^{N} \frac{\text{UB}_i^{(\alpha)} - \text{LB}_i^{(\alpha)}}{y_i} \tag{6.55}$$

where $y_{\max}$, $y_{\min}$ is the maximum and minimum value.

Comprehensive probabilistic metrics: The continuous ranked probability score (CRPS) is used in this study to evaluate the comprehensive performance of probabilistic prediction. The lower of CRPS representing the better estimation, the definition is as follows:

$$CRPS = \frac{1}{N} \sum_{i=1}^{N} \int_{-\infty}^{+\infty} \left[F(\hat{y}_i) - \text{II}_{\hat{y}_i > y_i} \right]^2 d\hat{y}_i \tag{6.56}$$

where $F(\hat{y}_i)$ is the cumulative distribution function.

Further, a metric simultaneously considering both stability and sharpness is proposed. The PI reliability and sharpness ratio (PIRSR) is defined as follows:

$$\text{PIRSR}^{(\alpha)} = \frac{\text{PINAW}^{(\alpha)} + \text{PIMWP}^{(\alpha)}}{\left(\text{PICP}^{(\alpha)}\right)^3} \tag{6.57}$$

Meanwhile, the average interval score (AIS) is introduced for the uncertainty analysis. When the observed value does not fall within the prediction interval, the AIS value is reduced by a penalty term calculating the formula as follows:

$$\pi_i^{(\alpha)} = \begin{cases} -2\alpha\left(\text{UB}_i^{(\alpha)} - \text{LB}_i^{(\alpha)}\right) - 4\left(\text{LB}_i^{(\alpha)} - y_i\right), \\ y_i < \text{LB}_i^{(\alpha)} \\ -2\alpha\left(\text{UB}_i^{(\alpha)} - \text{LB}_i^{(\alpha)}\right), \text{LB}_i^{(\alpha)} < y_i < \text{UB}_i^{(\alpha)} \\ -2\alpha\left(\text{UB}_i^{(\alpha)} - \text{LB}_i^{(\alpha)}\right) - 4\left(y_i - \text{UB}_i^{(\alpha)}\right), \\ y_i > \text{UB}_i^{(\alpha)} \end{cases} \tag{6.58}$$

$$\text{AIS} = \frac{1}{N} \sum_{i=1}^{N} \pi_i^{(\alpha)} \tag{6.59}$$

where $\pi_i^{(\alpha)}$ is the evaluation penalty term. The more interval deviating, the heavier penalty will be imposed.

6.4.3 Deterministic Prediction Analysis

In this section, the RMSE, MAPE, SDE, and LIB are considered as the evaluation indices of point prediction. For a typical electricity load, the target of point prediction is to generate accurate value with stable performance. Among them, the RMSE and MAPE describe the accuracy of the determined prediction while the SDE describes the stability of the result. LIB is proposed to evaluate the CPV problem in load forecasting.

To conduct the deterministic prediction analysis experiment, benchmark models are introduced to verify the validity and advantage of the proposed method. On the one hand, the simple sequence neural network model is designed as the benchmark model which is used to test the effectiveness of the proposed improvement to the basic form. The neural network hyperparameter is kept the same as the proposed method to avoid external interference. On the other hand, the widely recognized models for time-series prediction problems including random forest, RNN, and bidirectional LSTM (Bi-LSTM) are selected as the benchmark models that are used to verify the advantage of the proposed method. To reduce irrelevant factors, the experimental environment is consistent during the case study.

The deterministic prediction results of RMSE and MAPE are shown in Table 6.2. To comprehensively examine the seasonal difference of the electricity load, the indices of the models are, respectively, calculated in different seasons. As is shown, the RMSE and MAPE values achieved by the proposed framework are significantly lower than those of the counterparts in most cases, indicating a higher accuracy for electricity load forecasting. Compared with the selected models, the average promotion of the RMSE is 66.12%, 57.87%, 49.99%, and 21.52%, respectively, for other benchmark forecasting models. For the MAPE, the average promotion is correspondingly 60.53, 53.43, 41.26, and 18.46%. The performance of the forecasting model remains relatively stable across seasons, indicating the stability of model accuracy in seasonal differences. The instability of amplitude and abnormal changes caused by different external factors in the physical world results in the difference in predictability.

Specifically, to illustrate the forecasting performance intuitively, the load point curve between forecasting models from 17:00 August 16, 2016, to 12:00 September 3, 2016, is shown in Fig. 6.11. The curve of random forest is omitted due to large forecasting errors.

The advancement between the proposed framework and its basic forms is also verified. It is easy to find that the average promotion of the RMSE indices in electricity load cases obtained by the proposed framework is from 15.22 to 28.03% in comparison with the sequence neural network model. The progress is also achieved from 15.00 to 23.24% in MAPE, which further corroborated the accuracy improvement by integration of the attention mechanism and LCE approach.

The stability of determined prediction is also concern. For this reason, SDE is compared between models. The SDE is focused on the overall consistency of residuals, where a larger SDE reflects more significant instability of prediction. The bar

Table 6.2 The deterministic prediction results of RMSE and MAPE

Model	Metric	Period			
		Spring	Summer	Fall	Winter
Random Forest	RMSE	56.93	65.88	56.02	49.83
	MAPE	3.332%	3.634%	3.250%	2.948%
RNN	RMSE	43.84	49.26	45.46	45.29
	MAPE	2.755%	2.692%	2.725%	2.986%
Bi-LSTM	RMSE	39.87	41.22	37.76	36.04
	MAPE	2.267%	2.493%	2.108%	1.976%
Sequence	RMSE	22.81	25.49	25.66	24.74
	MAPE	1.560%	1.551%	1.600%	1.661%
Improved SANN	RMSE	17.96	18.34	20.18	20.98
	MAPE	1.250%	1.190%	1.343%	1.412%

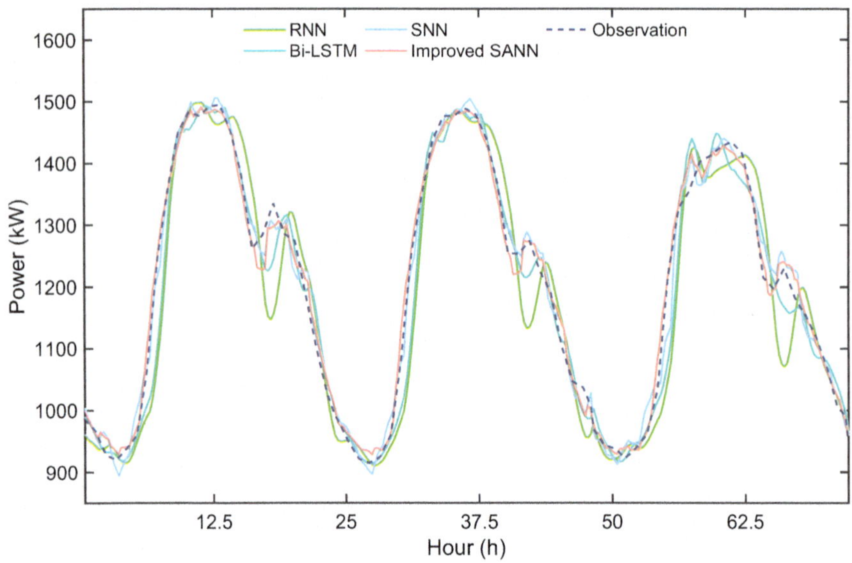

Fig. 6.11 The comparison of load point curve between forecasting models

graph shown in Fig. 6.12 demonstrates the comparison of SDE between prediction models. It can be concluded that the proposed framework tends to have a more stable result than its rivals in electricity load forecasting.

The defect of the CPV phenomenon in electricity load is emphasized in the previous section. The evaluation of the proposed liberty loss focuses on analyzing the similarity to the history of the load sequence. The LIB metric, which quantifies

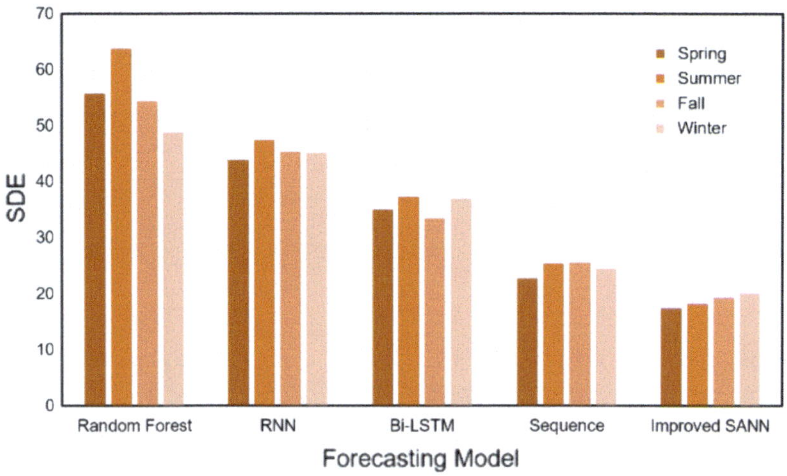

Fig. 6.12 The SDE metric comparison between prediction models

the extent of the CPV phenomenon, has been defined. A smaller LIB value indicates a heavier degree of CPV. As depicted in Table 6.3, the liberty loss consistently outperforms the MSE loss in all seasons when it comes to reducing value copying.

Specifically, the LIB value shows remarkable reductions from 52.22% to 71.66%, respectively. Moreover, Table 6.3 illustrates that mitigating the CPV phenomenon leads to an improvement in prediction error. The liberty loss yields lower RMSE and MAPE compared to the results obtained using MSE loss. Furthermore, a comparison of SDE indicates that forecasting stability also experiences simultaneous improvement. Consequently, mitigating the CPV phenomenon through liberty loss proves to be an effective approach for optimizing prediction performance.

To elaborate further, the occurrence of CPV is visually represented as a curve-shifting phenomenon. As depicted in Fig. 6.13, the curve of MSE loss reveals a

Table 6.3 The deterministic metrics between liberty and MSE loss

Metric	Loss	Period			
		Spring	Summer	Fall	Winter
LIB	MSE	1196.22	1593.63	1186.41	974.96
	Liberty	403.69	451.70	462.77	465.85
RMSE	MSE	37.54	35.12	38.77	40.69
	Liberty	17.96	18.34	20.18	20.98
MAPE	MSE	1.894%	1.975%	2.016%	2.089%
	Liberty	1.250%	1.190%	1.343%	1.412%
SDE	MSE	26.78	29.32	27.55	28.08
	Liberty	17.46	18.24	19.40	20.19

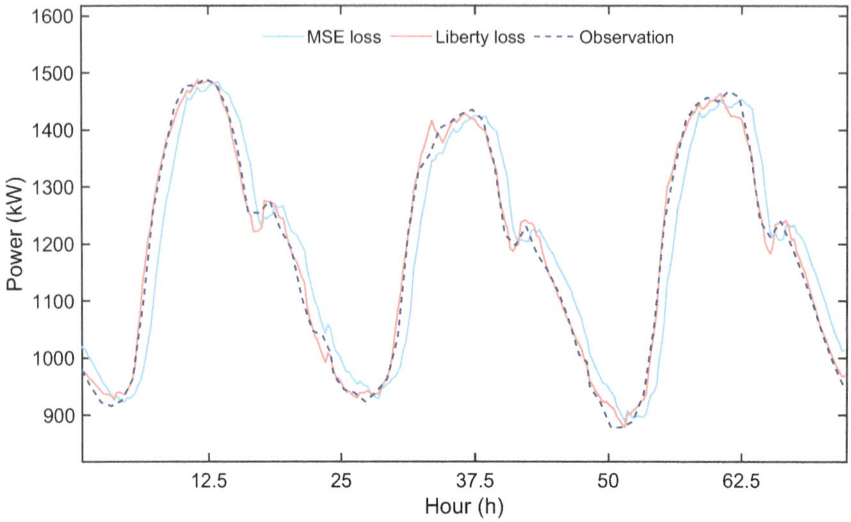

Fig. 6.13 The CPV illustration and losses comparison

noticeable lag in the CPV result. Typically, these lags predominantly occur within four timesteps. In real-world applications, such delays can be detrimental to system operations. However, the proposed model utilizing liberty loss effectively alleviates this phenomenon, offering valuable solutions for addressing practical challenges in load forecasting.

6.4.4 PI Stability and Sharpness Analysis

In this section, the PICP and ACE are considered as the evaluation indices of PI stability while PIMWP and PINAW are selected to evaluate sharpness. In general, PIRSR is compared in this part to give a balanced evaluation of PI stability and sharpness. For the following PI analysis result in the tables, PI nominal confidence level (PINC) α is set to 0.9, where quantile $\tau = [0.05, 0.95]$ is chosen as the prediction interval. Similarly, quantile $\tau = [0.10, 0.90]$ is chosen as the prediction interval of $\alpha = 0.8$.

The stability performance comparison of constructed PIs between probabilistic models is illustrated in Table 6.4. The coverage of PIs maintains an excellent level in all cases which is above 90% for the proposed framework. Meanwhile, the deviation measurement ACE is calculated simultaneously. The positive deviation is acceptable while the larger negative ACE reflects the substandard of PI stability.

It is found that all ACEs of the proposed framework are positive while the smallest deviation reaches 0.0187. The coverage probability and the sharpness are a pair of contradictions. In very few cases, the proposed framework slightly falls behind in

Table 6.4 Stability performance comparison of constructed PIs of probabilistic models

Model	Metric	Period			
		Spring	Summer	Fall	Winter
Random Forest	PICP	0.8188	0.8375	0.8104	0.7889
	ACE	−0.0813	−0.0625	−0.0896	−0.1111
RNN	PICP	0.9438	0.8993	0.9250	0.9410
	ACE	0.0437	−0.0007	0.0250	0.0410
Bi-LSTM	PICP	0.8731	0.8955	0.8647	0.9021
	ACE	−0.0269	−0.0045	−0.0353	0.0021
Sequence	PICP	0.8389	0.8472	0.8215	0.8604
	ACE	−0.0611	−0.0528	−0.0785	−0.0396
Improved SANN	PICP	0.9451	0.9188	0.9507	0.9528
	ACE	0.0451	0.0187	0.0507	0.0528

PICP. This is because the counterpart chose to sacrifice the sharpness, which will be analyzed in the following part.

Furthermore, the sharpness performance comparison of constructed PIs between probabilistic models is presented in Table 6.5. It can be concluded that the PINAW of the proposed framework is better than the other models in most cases, which shows excellent sharpness performance. On the one hand, the sharpness of PI is necessary since the excess wide interval is lacking in value for practical system operators. However, the pursuit of a narrow load prediction boundary would lead to a drop in coverage probability.

Therefore, the PIRSR is compared to give a balancing estimation. The result shows that the proposed framework achieves the best performance in balancing stability and sharpness. The practical probabilistic prediction curve is shown in Fig. 6.14, and the best performing rival is presented for comparison. It is found that the conclusion is consistent with the above analysis, indicating the effectiveness of generating high reliability and sharpness PI for electricity load.

6.4.5 Comprehensive Probabilistic Prediction and Multistep Discussion

This section focuses on the analysis of comprehensive probabilistic prediction and multistep prediction. To ensure a robust evaluation framework, two indicators are used: the CRPS serves as a positive indicator for probability distribution assessment, while the AIS serves as a negative indicator, reflecting penalties for inaccurate estimation.

The AIS is utilized to assess the deviation of predictions, leading to significant penalties for inaccurate estimations and distinguishing suitable probabilistic

Table 6.5 Sharpness performance comparison of constructed PIs of probabilistic models

Model	Metric	Period				Average PIRSR
		Spring	Summer	Fall	Winter	
Random forest	PIMWP	0.2369	0.2361	0.2361	0.2372	0.9531
	PINAW	0.2740	0.2878	0.2743	0.2689	
	PIRSR	0.9309	0.8918	0.9588	1.0308	
RNN	PIMWP	0.1612	0.1487	0.1543	0.1645	0.4280
	PINAW	0.1861	0.1823	0.1806	0.1859	
	PIRSR	0.4131	0.4550	0.4231	0.4207	
Bi-LSTM	PIMWP	0.1093	0.1145	0.0937	0.1287	0.3564
	PINAW	0.1308	0.1259	0.1442	0.1371	
	PIRSR	0.3608	0.3348	0.3678	0.3621	
Sequence	PIMWP	0.0689	0.0679	0.0665	0.0674	0.2459
	PINAW	0.0794	0.0824	0.0774	0.0764	
	PIRSR	0.2512	0.2472	0.2595	0.2258	
Improved SANN	PIMWP	0.0825	0.0778	0.0844	0.0882	0.2115
	PINAW	0.0923	0.0915	0.0945	0.0955	
	PIRSR	0.2070	0.2183	0.2083	0.2124	

results. Table 6.6 displays the average AIS values for load results across different models. The proposed framework incurs the lightest penalties in all cases, indicating the high-quality probability prediction it achieves. This conclusion is further corroborated by the CRPS results in Fig. 6.15, which also show the lowest average CRPS for the proposed method. The result shows that the proposed method achieves a maximum improvement of 22.92% when compared to benchmarks. In summary, the improved SANN outperforms other competing models in comprehensive probabilistic prediction for electricity load.

Furthermore, to evaluate the performance of the proposed probabilistic forecasting model in a longer forecast time, multistep forecasting is conducted. To be specific, the forecast time scale is set from 15 min to 4 h. The above models are trained to generate multistep forecasting results to meet the demand for longer forecasts. Four metrics are selected to measure the multistep performance, including RMSE, MAPE, PIMWP, and PICP.

A comparison of probabilistic load forecasting models in different steps is demonstrated in Fig. 6.16. It can be concluded from the figure that the proposed framework remains the advantage in both deterministic and probabilistic forecasting among different time scales. As the forecast time scale increases, the proposed framework achieves less performance loss which presents a gentler slope in the graph. In PIMWP and PICP, the curve of random forest is omitted due to a large deviation of the metric. The above results indicate that the proposed load forecasting framework is capable of longer forecasts and has the potential for higher power system operational requirements.

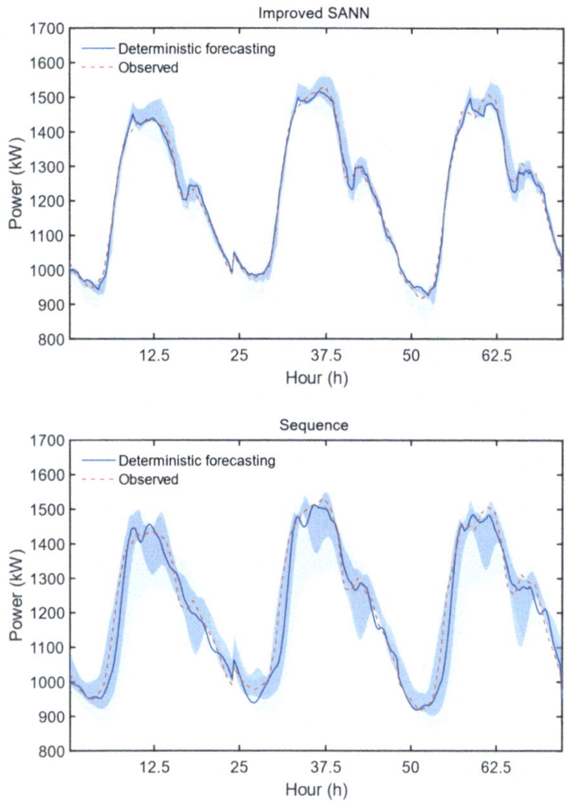

Fig. 6.14 The probabilistic forecasting of load by different models

Table 6.6 AIS comparison of probabilistic load forecasting models

Model	Period			
	Spring	Summer	Fall	Winter
Random forest	−508.62	−535.08	−508.83	−506.32
RNN	−337.92	−333.75	−329.27	−336.65
Bi-LSTM	−298.43	−307.51	−294.68	−304.13
Sequence	224.37	−233.68	−223.06	−216.04
Improved SANN	−168.28	−168.13	−173.86	−173.77

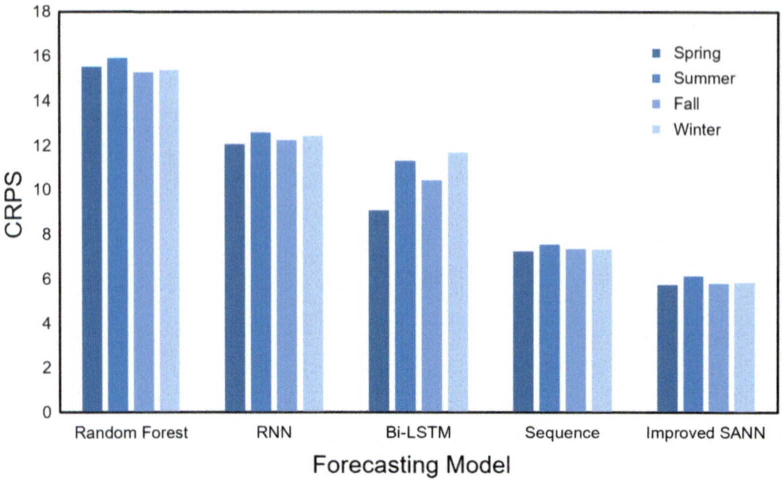

Fig. 6.15 CRPS comparison of probabilistic load forecasting models

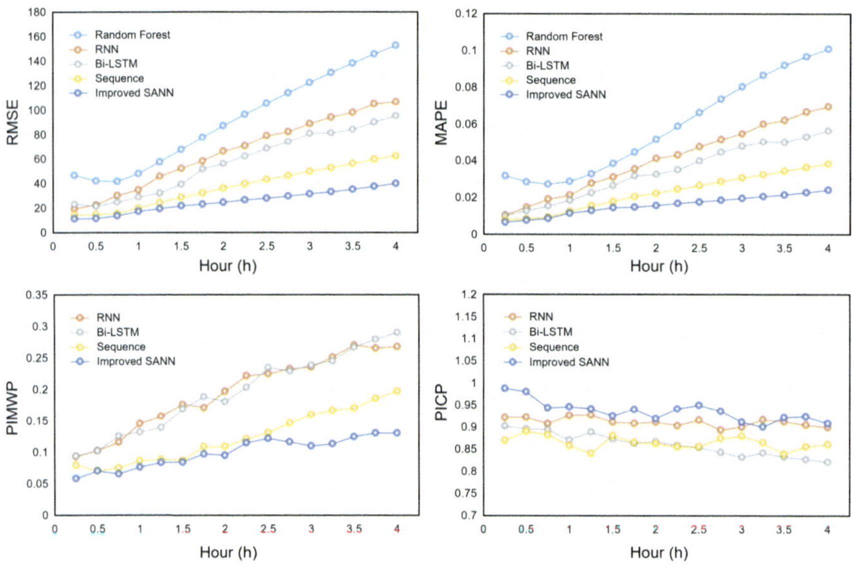

Fig. 6.16 Multistep comparison of probabilistic load forecasting models

6.5 Conclusion

In this chapter, an improved SANN-based hybrid load probabilistic forecasting framework for the increasing demand of smart grid operation is proposed. Considering the existing problem proposed in the first section, the framework made improvements in multiple aspects. The LCE approach is designed to identify the most influential features while extracting deeper patterns of load. Aiming at generating probability distributions for future load values, offering a more comprehensive understanding of potential outcomes and associated uncertainties, kernel density estimation is integrated with improved SANN forming a load forecasting framework. Meanwhile, the enhanced loss function effectively mitigates the CPV problem while reducing the forecasting error for load value. The case study validates that the proposed framework is capable of providing probabilistic forecasting with higher accuracy, stability, and reliability. Future work will focus on day-ahead forecasting and more detailed modeling for load or renewable energies.

References

1. C. Wan, W. Qian, C. Zhao, Y. Song, G. Yang, Probabilistic forecasting based sizing and control of hybrid energy storage for wind power smoothing. IEEE Trans. Sustain. Energy **12**(4), 1841–1852 (2021)
2. IEA, Consumption by household appliances and plug loads by region in the Sustainable Development Scenario (IEA, Paris, June 2020)
3. Y. Li, S. He, Y. Li, L. Ge, S. Lou, Z. Zeng, Probabilistic charging power forecast of EVCS: reinforcement learning assisted deep learning approach. IEEE Trans. Intell. Veh. **8**(1), 344–357 (2023)
4. Z. Yi et al., An improved two-stage deep reinforcement learning approach for regulation service disaggregation in a virtual power plant. IEEE Trans. Smart Grid **13**(4), 2844–2858 (2022)
5. C. Wang, Y. Wang, Z. Ding, T. Zheng, J. Hu, K. Zhang, A transformer-based method of multienergy load forecasting in integrated energy system. IEEE Trans. Smart Grid **13**(4), 2703–2714 (2022)
6. T. Hu, H. Ma, H. Liu, H. Sun, K. Liu, Self-attention-based machine theory of mind for electric vehicle charging demand forecast. IEEE Trans. Ind. Inform. **18**(11), 8191–8202 (2022)
7. J.-F. Toubeau, J. Bottieau, F. Vallée, Z. De Grève, Deep learning-based multivariate probabilistic forecasting for short-term scheduling in power markets. IEEE Trans. Power Syst. **34**(2), 1203–1215 (2019)
8. A. Azeem, I. Ismail, S.M. Jameel, V.R. Harindran, Electrical load forecasting models for different generation modalities: a review. IEEE Access **9**, 142239–142263 (2021)
9. Y. Wang, Q. Chen, M. Sun, C. Kang, Q. Xia, An ensemble forecasting method for the aggregated load with subprofiles. IEEE Trans. Smart Grid **9**(4), 3906–3908 (2018)
10. Y. Wang, Q. Chen, N. Zhang, Y. Wang, Conditional residual modeling for probabilistic load forecasting. IEEE Trans. Power Syst. **33**(6), 7327–7330 (2018)
11. K. Wang, Y. Zhang, F. Lin, J. Wang, M. Zhu, Nonparametric probabilistic forecasting for wind power generation using quadratic spline quantile function and autoregressive recurrent neural network. IEEE Trans. Sustain. Energy **13**(4), 1930–1943 (2022)
12. F. Wu, R. Jing, X.-P. Zhang, F. Wang, Y. Bao, A combined method of improved grey BP neural network and MEEMD-ARIMA for day-ahead wave energy forecast. IEEE Trans. Sustain. Energy **12**(4), 2404–2412 (2021)

13. J. Browell, M. Fasiolo, Probabilistic forecasting of regional net-load with conditional extremes and gridded NWP. IEEE Trans. Smart Grid **12**(6), 5011–5019 (2021)
14. L. Sang, Y. Xu, H. Long, Q. Hu, H. Sun, Electricity price prediction for energy storage system arbitrage: a decision-focused approach. IEEE Trans. Smart Grid **13**(4), 2822–2832 (2022)
15. Y. Wang, N. Gao, G. Hug, Personalized federated learning for individual consumer load forecasting. CSEE J. Power Energy Syst. **9**(1), 326–330 (2023)
16. L. Yang, M. He, J. Zhang, V. Vittal, Support-vector-machine-enhanced markov model for short-term wind power forecast. IEEE Trans. Sustain. Energy **6**(3), 791–799 (2015)
17. Y. Liu, J. Wang, Transfer learning based multi-layer extreme learning machine for probabilistic wind power forecasting. Appl. Energy **312**, 118729 (2022)
18. M. Zou et al., Bayesian CNN-BiLSTM and Vine-GMCM based probabilistic forecasting of hour-ahead wind farm power outputs. IEEE Trans. Sustain. Energy **13**(2), 1169–1187 (2022)
19. Y. Guo et al., BiLSTM multitask learning-based combined load forecasting considering the loads coupling relationship for multienergy system. IEEE Trans. Smart Grid **13**(5), 3481–3492 (2022)
20. K. Doubleday, S. Jascourt, W. Kleiber, B.-M. Hodge, Probabilistic solar power forecasting using bayesian model averaging. IEEE Trans. Sustain. Energy **12**(1), 325–337 (2021)
21. P. Zhao, D. Cao, Y. Wang, Z. Chen, W. Hu, Gaussian process-aided transfer learning for probabilistic load forecasting against anomalous events. IEEE Trans. Power Syst. **38**(3), 2962–2965 (2023)
22. H. Liu et al., Short-term wind power interval prediction method using VMD-RFG and att-GRU. Energy **251**, 123807 (2022)
23. Z. Zhang et al., Wind speed forecasting based on Quantile Regression Minimal Gated Memory Network and Kernel Density Estimation. Energy Convers. Manag. **196**, 1395–1409 (2019)

Chapter 7
Optimal Communication Network Design of Wide Area Measurement System

Yuan Yao and Yinliang Xu

7.1 Introduction

7.1.1 Background

A microgrid is essentially a cyber-physical system in which the communication network enables data exchange among distributed energy resources (DERs) for monitoring and control purposes. Such communication structures should be optimized to mitigate communication delays/failures and hence, increase the overall reliability. Wide area measurement system (WAMS) has been widely applied on the smart grid and microgrids during the past few decades [1, 2]. It consists of the phasor measurement units (PMUs), the phasor data concentrator (PDC), and a communication network connecting them. The function of the WAMS communication network is to transfer the measurement data of the system operation state collected by PMUs to the PDC. The measurement data received by the PDC will then be used in the WAMS application [3], such as state estimation [4] and stability control [5]. With the increasing number of PMUs, the volume of transferred data in the communication network increases. It aggravates the complexity of data transmission and increases the end-to-end (ETE) time delay of the data experienced, thus affecting the data completeness received by the PDC and the performance of the WAMS applications [6]. To minimize the ETE delay and guarantee the reliability of transferred data, it is vital to propose an approach that can effectively select the optimal PDC placement and the optimal communication paths in WAMS communication network.

Y. Yao · Y. Xu (✉)
Tsinghua Shenzhen International Graduate School, Tsinghua-Berkeley Shenzhen Institute, Tsinghua University, Shenzhen 518055, P. R. China
e-mail: xu.yinliang@sz.tsinghua.edu.cn

Y. Yao
e-mail: yao-y21@mails.tsinghua.edu.cn

J. Hu (ed.), *Community Energy and Microgrids*, Green Energy and Technology,
https://doi.org/10.1007/978-981-97-6297-2_7

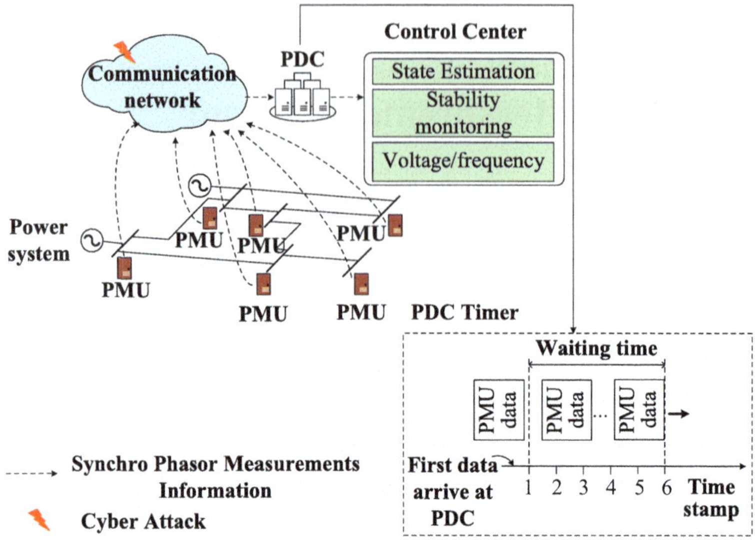

Fig. 7.1 The overall structure of WAMS

At the same time, cyber-attacks are becoming a problem that needs to be considered and solved in the WAMS communication network [7]. It may bring in a large amount of redundant data to occupy some communication paths, thus causing a longer ETE delay for the PMU data suffered and the PDC to possibly discard it [8]. Therefore, the proposed approach needs to meet the real-time requirements for dynamic path reconstruction. The overall structure is shown in Fig. 7.1.

7.1.2 Literature Review

The path planning problem relates to a complex decision-making process, and some of them have been proven to be NP-hard [9, 10]. In recent years, there have been several researches focusing on the optimal communication path design between PMUs and the PDC. According to the different objectives, the existing research can be divided into two categories:

(1) Minimize the construction cost of the communication path. A BILP model to determine the optimal placement of PMUs, the PDC, and the communication paths between them is proposed in [11]. Reference [12, 13] leverages the Dijkstra algorithm to optimize the PDC placement and design the WAMS communication paths. These methods actually consider the communication length minimization, but ignore the influence of ETE delay from the communication system to the power system.

(2) Minimize the propagation and transmission delay of the data experienced. In [14], the genetic algorithm (GA)-based approach aims to minimize the propagation delay experienced by the PMU data. In [15], the transmission delay is also taken into account. To explore the impact of ETE delay on the WAMS applications, [15] compares the accuracy of the voltage phasor estimation under different data arrival rates. Although the method in [15] considers the propagation and transmission delay, it ignores temporal coefficients that are related to the queuing delay, which may lead to an error in ETE delay calculation and influence the path planning decision.

The researchers further considered the scenarios where cyber-attacks appear in WAMS communication systems and analyzed the impact of cyber-attacks on WAMS applications. The cyber-attacks generally consist of Denial of Service (DoS) and data spoofing [16]. The DoS attacks generate massive redundant data traffic to the target PMUs [17] or communication links [15] so that they cannot operate properly. A communication path reconstruction strategy is given in [15] when the DoS attacks the communication links. The data spoofing refers to the wrong data injection into PMUs and duplication of the past measurements of PMUs [18]. The solutions are mainly to develop robust filters to identify deceptive data to improve the immunity of the WAMS applications [19, 20]. In this chapter, our focus is the path of reconstruction under the DoS attacks. Specifically, it considers the transmission delay increase and the complete breakage of some communication links.

The real-time requirement of path reconfiguration methods is critical, and it is often more challenging to achieve this through centralized algorithms [11–13]. Reference [21] first applies a distributed approach, i.e., the consensus algorithm, to deal with the shortest path problem. It only requires the local network node to exchange information with its neighbors and reach an agreement on a certain quantity. Therefore, it is efficient to meet the real-time requirements and scalable to large systems. The consensus-based algorithms are then extended to other applications, such as P2P energy trading [22] and path design for uncrewed surface vessels [23]. This technique provides a new perspective on the dynamic path planning between PMU and PDC in WAMS.

7.1.3 Contributions

In this chapter, a distributed bias min-consensus (DBMC)-based approach is proposed to optimize the WAMS communication network. The objective is to minimize the ETE delay and maximize the data arrival rate received by the PDC under normal conditions and cyber-attacks. Meanwhile, the influence of the ETE delay on the state estimation is explored. The performance of the proposed approach is validated through the IEEE 39-bus and IEEE 118-bus test systems.

The major contributions are summarized as follows.

1. This chapter proposes a distributed DBMC algorithm to (re)construct paths between PMUs and the PDC to address the data incompleteness problem from the cyber-physical perspective.
2. Compared with the existing centralized algorithms [11–14], the proposed DBMC algorithm only requires information exchange between adjacent nodes and performs the local computation in each iteration, thus reducing the communication and computation burden. At the same time, it can guarantee the global optimal solution and meet the real-time requirements of dynamic path reconstruction under cyber-attacks.
3. A delay detection module is added. It further considers the spatial–temporal relationship for the queuing delay of each path and is more relevant to the practical situation in path planning.

7.2 Problem Formulation

7.2.1 ETE Delay Calculation

The ETE delay experienced on link (i, j) can be calculated as

$$\tau_{ij,p} = \tau_{ij}^{tran} + \tau_{ij}^{prop} + \tau_{ij,p}^{queu} \tag{7.1}$$

where $\tau^{tran} = M \big/ R_{ij}$ is the transmission delay, $\tau^{prop} = d_{ij} \big/ v$ is the propagation delay, τ^{queu} is the queuing delay, M is the data packet size, R_{ij} is the bandwidth of the communication link, d_{ij} is the length of the link, and v is the propagation speed of the transferred data packet.

The queuing delay τ^{queu}, which is difficult to get in advance, is dependent on the communication network's traffic condition and the queuing mechanisms. Some research leverages the average queuing delay [15] or restricts the maximum amount of data allowed by each link [24] to estimate the queuing delay, which is not accurate enough.

In this chapter, the First-in-First-out (FIFO) is considered as the queuing mechanism. Take two PMUs transferring data to one PDC as an example. Denote τ_1 and τ_2 as the ETE delay experienced by the two PMU data. If $\tau_1 > \tau_2$, the queuing delay of data from PMU 1 can be calculated as

$$\tau_{l,1}^{queu} = \begin{cases} \tau_2 + \tau_3 - \tau_1 \, (\tau_1 < (\tau_2 + \tau_3)) \\ 0 \, (\tau_1 \geq (\tau_2 + \tau_3)) \end{cases} \tag{7.2}$$

The Simulink in MATLAB is applied to model and simulate the queuing delay that may exist in the communication link. Let the server time in the block Server mimic the transmission and propagation delay experienced by the PMU data. The

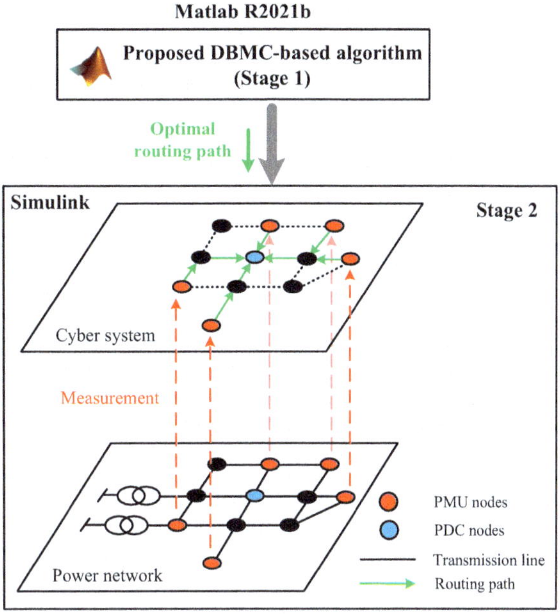

Fig. 7.2 The designed framework in the manuscript

FIFO block is used to simulate the queue process. The scopes in the FIFO and PDC block are used to record the waiting time of each data in this queue (i.e., the queuing delay) and the data arrival time. The total framework is shown in Fig. 7.2, and the details of the simulation platform are as follows.

In Fig. 7.2, the proposed DBMC-based algorithm is implemented in MATLAB R2021b. The related process, including the PMU data sending, the ETE delay suffered by the data of PMUs to the PDC, and the data reception of PDC, is simulated in the Simulink platform. The details of the simulation platform are as follows:

1. *The PMU models* (*The PMU data sending*)

The Entity Generator block is used to simulate the process of PMU sending data periodically by generating periodic signals. (1) Its generating rate is consistent with that of the PMU, which is set to 30 ms in this chapter. (2) The ETE delay (without the queuing delay) suffered on each link along the path is set in this block, which will be recognized by the link it transferred. The logo of *the Entity Generator block* is given in Fig. 7.3.

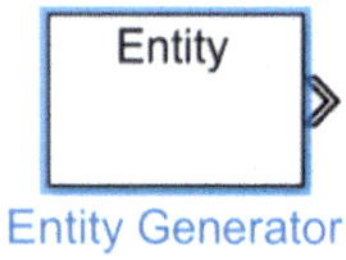

Fig. 7.3 The logo of the Entity Generator block

Fig. 7.4 The logo of the Entity Terminator block

2. *The PDC models (the data reception of PDC)*

The Entity Terminator block is applied to simulate the data reception process of the PDC. By selecting the option "Number of entities arrived, a" and adding a Scope in this port, the arrival time of each PMU data can be recorded, that is, the ETE delay that they experienced. Then, the data arrival rate under a given PDC waiting time can be obtained. The logo of the Entity Terminator block is given in Fig. 7.4.

3. *The links models (the transmission and propagation delay suffered by the data of PMUs to the PDC)*

The links of each path are modeled by the Entity Server block. Each Entity Server block corresponds to a specific link. As mentioned in the Entity Generator block, the ETE delay (without the queueing delay) suffered on each link will be performed by the Entity Server blocks. "1" refers to the capacity of the link, which means it cannot transfer multiple data at the same time. The logo of the Entity Server blocks is given in Fig. 7.5.

4. *The queueing models (the queuing delay suffered by the data of PMUs to the PDC)*

In our manuscript, the queuing mechanism selects the First-in-First-out (FIFO). Therefore, the FIFO Entity Queue block in the Simulink is considered. By selecting the option "Number of entities in block, n" and adding a Scope in this port, the arrival and leaving time in this block of each PMU data can be recorded. That is, the queuing time suffered on this link of each PMU data is obtained. The logo of the FIFO Entity Queue block is given in Fig. 7.6.

By now, all the elements related to the cyber system are listed in the Simulink. Continuing to illustrate with the example above, the corresponding model in Simulink is shown in Fig. 7.7.

Take the IEEE-30 bus system in our case study as an example. Node 6 is selected as the optimal PDC placement. The corresponding paths connecting PMUs and the PDC are shown in Fig. 7.8.

Fig. 7.5 The logo of the Entity Server block

Fig. 7.6 The logo of the FIFO Entity Queue block

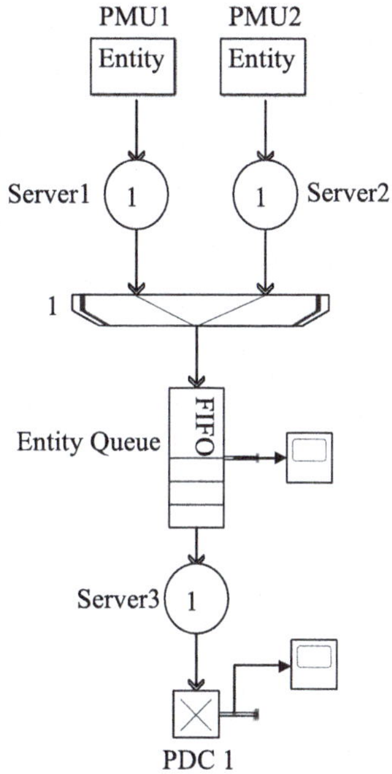

Fig. 7.7 A simple queuing model developed in Simulink

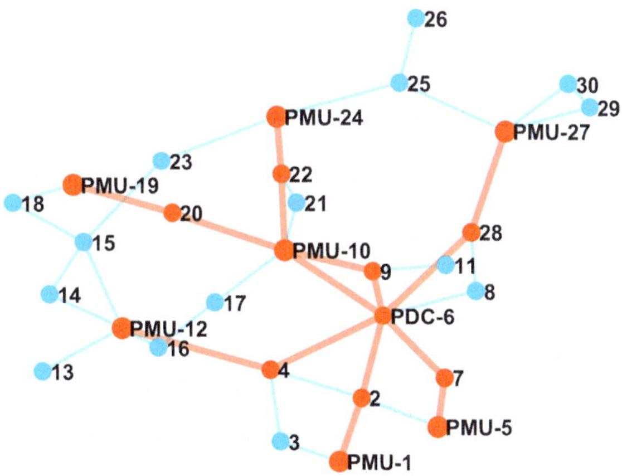

Fig. 7.8 The optimal paths planning in IEEE-30 bus system

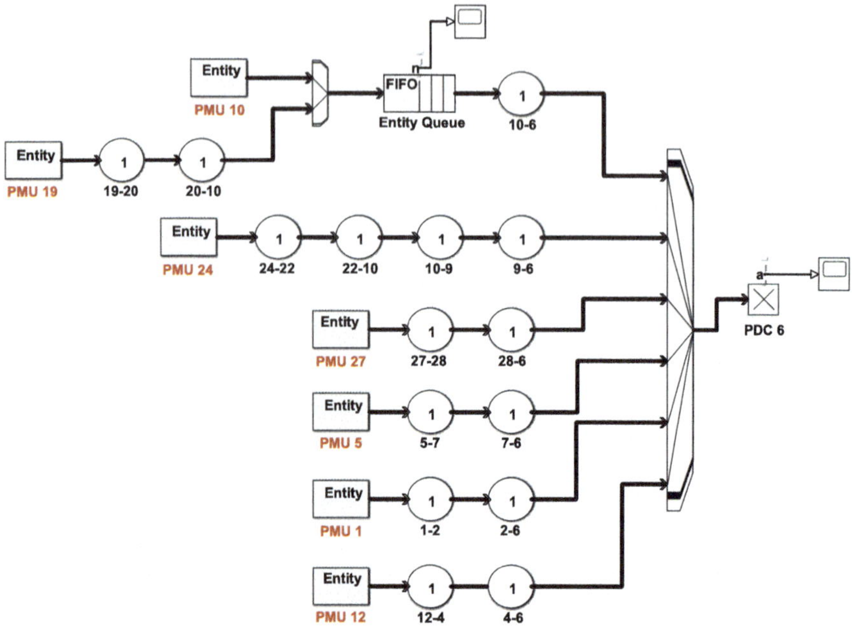

Fig. 7.9 The communication network design in Simulink

Correspondingly, the established path topology in Simulink is consistent with that in Fig. 7.9. Note that the path construction process in Simulink is automatically formulated by the procedure coding in MATLAB 2021b.

In the Simulink platform, the simulation duration is set to 30 ms, that is, one data sending cycle. The arrival time of PMU data (i.e., the ETE delay suffered by each PMU data) received by the PDC obtained in Simulink and our algorithm are shown in Fig. 7.10 and Table 7.1.

In Fig. 7.10, it can be found that the data arrival time obtained from the proposed method is consistent with the results generated by Simulink, which verifies the effectiveness of the proposed method.

Also, by checking *the number of entities in the block* of the FIFO module, the queuing delay experienced by the PMU data on links can be determined. In this case, the data of PMU 10 and 19 experience the same link (10, 6). The number of entities in the block of the FIFO module corresponding to this link is exported in Fig. 7.11.

In Fig. 7.11, it can be seen that the data of PMU 10 arrives first, then leaves immediately in the first period. After about 6.5 ms, the data of PMU 19 arrives. Since the first arrived data has already passed through the link (10, 6) at this time, the data also does not experience queuing delay.

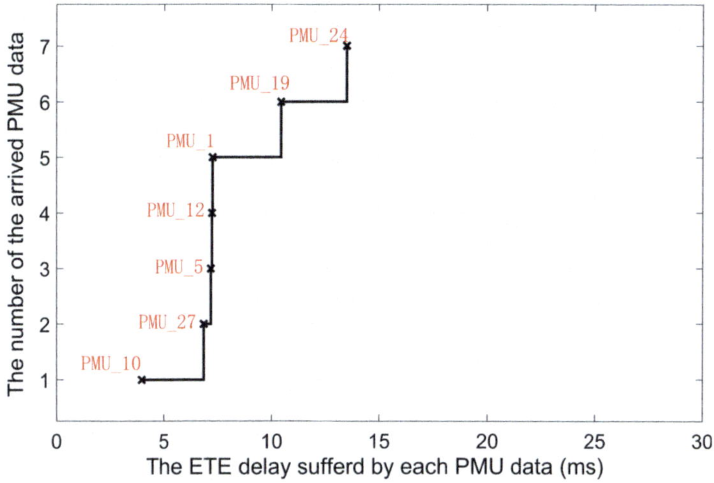

Fig. 7.10 The data arrival time of each PMU data obtained in Simulink

Table 7.1 The data arrival time of each PMU data to the PDC

PMU	Data arrival time (*ms*)	PMU	Data arrival time (*ms*)
10	3.91	1	7.29
27	6.87	19	10.44
5	7.20	24	13.51
12	7.26		

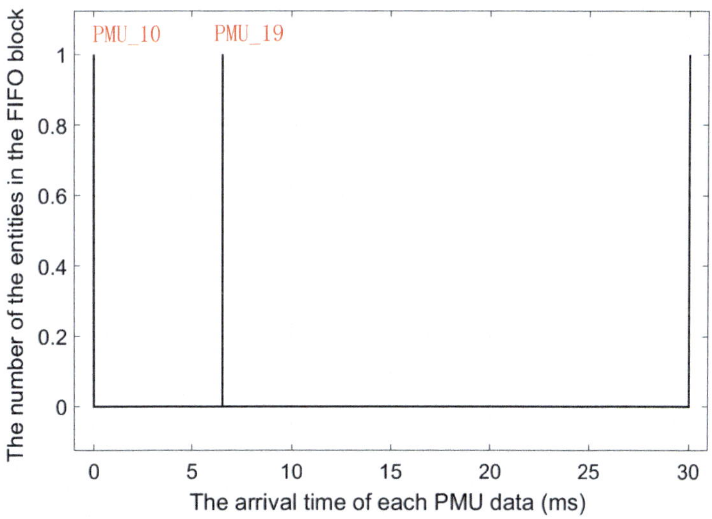

Fig. 7.11 The number of the entities in the FIFO blocks

7.2.2 PMU-Based State Estimation in WAMS

Since the performance of state estimation is highly affected by the completeness of measurement data received by the PDC, the influence of the ETE delay on the state estimation is explored in this chapter. Each PMU captures the measurement of the voltage phasor $\dot{u}_i$ and line current $\dot{I}_{ij}$ that directly connects with it.

To linearize the nonlinear measurement equation, an auxiliary variable is introduced and given as

$$\boldsymbol{x}_a = [\mathrm{Re}(\dot{u}_i), \mathrm{Im}(\dot{u}_i)]^T \tag{7.3}$$

where Re(·) and Im(·) represent the real and imaginary parts of the phasor. The measurement $z = [\mathrm{Re}(\dot{u}_i), \mathrm{Im}(\dot{u}_i), \mathrm{Re}(\dot{I}_{ij}^m), \mathrm{Im}(\dot{I}_{ij}^m)]^T$. The linearized measurement equations are detailed in [25].

Then, the weighted least squares (WLS)-based SE model is developed as

$$\begin{cases} \min[z - h(\boldsymbol{x}_a)]^T \boldsymbol{R}^{-1}[z - h(\boldsymbol{x}_a)] \\ s.t. z = \boldsymbol{H}\boldsymbol{x}_a + \boldsymbol{r}. \end{cases} \tag{7.4}$$

where $\boldsymbol{r}$ represents the measurement error with Gaussian distribution. $\boldsymbol{R} = Cov(\boldsymbol{r}) = diag\{\sigma_1^2, \sigma_2^2, \cdots, \sigma_m^2\}$. σ_i represents the standard deviation of each measurement error. $\boldsymbol{H}$ is the Jacobian matrix with constant values.

The estimated value of the auxiliary variables can be directly obtained as

$$\tilde{\boldsymbol{x}}_a = (\boldsymbol{H}^T\boldsymbol{R}^{-1}\boldsymbol{H})^{-1}\boldsymbol{H}^T\boldsymbol{R}^{-1}z \tag{7.5}$$

By nonlinear transformation, the state variables $\boldsymbol{x} = [\boldsymbol{u}, \theta]^T$ can be obtained, given as

$$\begin{cases} u_i^2 = \mathrm{Re}(\dot{u}_i)^2 + \mathrm{Im}(\dot{u}_i)^2 \\ \theta_i = \arctan(\mathrm{Re}(\dot{u}_i)/\mathrm{Im}(\dot{u}_i)) \end{cases} \tag{7.6}$$

7.3 Bias Min-consensus-Based Algorithm for Optimal Path Planning

The goal of this chapter is to determine the optimal paths between PMUs and the PDC to maximize the completeness of the data received by the PDC. To fully address the problem, three factors need to be considered: (1) The PMUs configuration; (2) The path planning between each PMU to the PDC candidate; (3) The optimal PDC placement. The PMUs configuration and the optimal PDC placement is determined during the planning stage and will not change in subsequent online operation.

7.3.1 PMUs Configuration

This chapter aims to minimize the number of PMUs while ensuring the system's observability and considers the effect of zero-injection nodes to obtain the configuration of PMUs. The observable network refers to the state of each network node can be determined by the measurements provided by the configured PMUs. Limited by the high cost of PMU, it is unrealistic to configure PMU at each node of the whole network. This chapter aims to minimize the installation cost of PMUs as well as ensure the networks' observability. The model development is given as follows:

1. A PMU installed on a node can provide the status of that node (the voltage magnitude and phase angle) and the branches' current phasor connected with the node. When a PMU is placed on a node, all the nodes connected to it become observable. So, the OPP problem can be expressed as

$$\begin{aligned} &\min \sum_{i=1}^{n} w_i x_i \\ &s.t. \sum_{j=1}^{n} a_{i,j} x_j + x_i \geq 1\ (\forall i \in V) \end{aligned} \tag{7.7}$$

where $\boldsymbol{x} = [x_1, x_2, \cdots x_n]^T$ is the decision variable. $x_i = 1$ if node i configures with a PMU, otherwise, $x_i = 0$. w_i is the PMU installation cost at node i. $V \in \mathbb{R}^n$ is the set of nodes in the network. $a_{i,j}$ is the element of the adjacent matrix $\boldsymbol{A} \in \mathbb{R}^{n \times n}$, i.e.,

$$a_{i,j} = \begin{cases} 1\ (\text{if}\, i = j) \\ 1\ (\text{if}\, i\, \text{and}\, j\, \text{are connected}) \\ 0\ (\text{otherwise}) \end{cases} \tag{7.8}$$

2. Modeling of the zero-injection nodes

Zero-injection nodes correspond to the nodes with no current injected. The states of zero-injection buses and their adjacent nodes can be obtained by applying Kirchhoff's law.

Given an IEEE 9-bus system as an example, where there exist three zero-injection nodes including nodes 4 and 6. Without considering the effects of the zero-injection nodes, there are four PMUs that ensure the overall system's observability shown in Fig. 7.12a. If we consider the influence of the zero-injection node, there are only two PMUs that can guarantee the systems' observability, as given in Fig. 7.12b. The reasons are as: for the zero-injection node 4, since nodes 5 and 9 are observable, the voltage $\dot{u}_5$, $\dot{u}_9$ and the branch current $\dot{I}_{54}$, $\dot{I}_{94}$ are known. Therefore, the voltage of node 4 can be calculated by

$$\dot{u}_4 = \dot{u}_5 - \dot{I}_{54}Z_{54} = \dot{u}_9 - \dot{I}_{94}Z_{94} \tag{7.9}$$

where Z_{54} and Z_{94} are the branch impedance.

The node 4 is a zero-injection node. By applying the Kirchhoff's current law (KCL), the branch current $\dot{I}_{41}$ can be calculated as

$$\dot{I}_{41} = -(\dot{I}_{54} + \dot{I}_{94}) \tag{7.10}$$

Then, the voltage $\dot{u}_3$ can be obtained:

$$\dot{u}_1 = \dot{u}_4 - \dot{I}_{41}Z_{41} \tag{7.11}$$

Fig. 7.12 Influence of zero-injection nodes in OPP problem

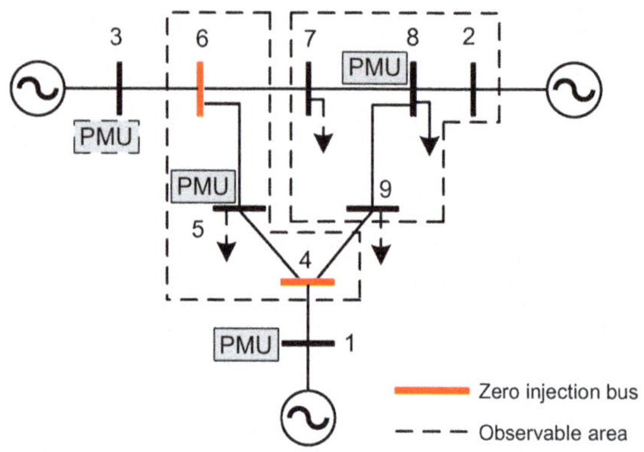

(a) PMUs configuration without considering zero-injection nodes

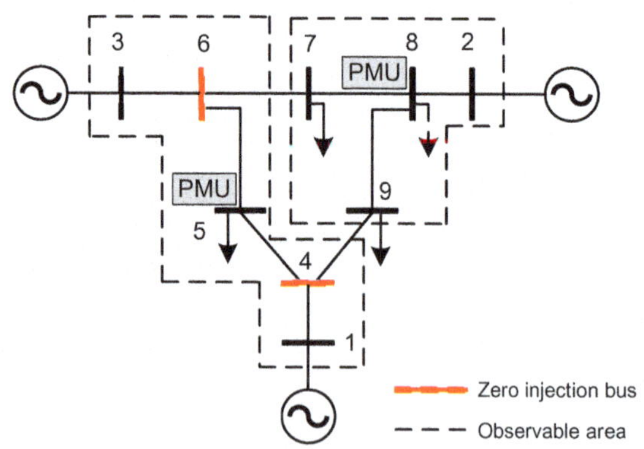

(b) PMUs configuration considering zero-injection nodes

Therefore, node 1 is also observable without installing PMU on it. The same conclusion applies to node 6 and node 3. So, if zero-injection nodes are also modeled in the OPP problem, the total number of PMUs can further be reduced [26].

The way to model the zero-injection nodes in this chapter is to impose additional variables and constraints on them. That is, for a zero-injection node and its adjacent nodes (the total number is N), when N-1 nodes are all observable, the remaining node is also observable. Therefore, the OPP model (7) can be further derived as

$$
\begin{aligned}
&\min \sum_{i=1}^{n} w_i x_i \\
&s.t. \sum_{j=1}^{n} a_{i,j} x_j + x_i - u_i \geq 0\,(\forall i \in V) \\
&\sum_{j=1}^{n} a_{i,j} u_j \geq B_i - 1\,(\forall i \in V_{nz})
\end{aligned}
\tag{7.12}
$$

where $V_{nz} \in \mathbb{R}^{n_k}$ represents the set of zero-injection nodes. B_i is the number of adjacent nodes of the ith zero-injection node (including the zero node itself). u_j is the jth element of $\boldsymbol{u} \in \mathbb{R}^n$, i.e.,

$$
u_j = \begin{cases} 1\,(\text{if}\, j \notin B_1 \cup B_2 \cup \cdots \cup B_{n_k}) \\ u_j(otherwise) \end{cases}
\tag{7.13}
$$

7.3.2 Optimal Paths Planning Between PMUs and the PDC

The proposed optimal path planning approach is applied for the optimal path (re)planning between PMUs and the PDC during the off-line planning and under cyber-attacks in online operations. Since the queuing delay cannot be determined before path planning like the physical distance in [21], a detection unit is added to check the queuing delay in the existing tree and decide whether the current paths need to be re-routed. Details are provided as follows.

Let $y_i(k)$ represents the ETE delay from node i to the PDC node without considering the queuing delay. The updating protocol in each iteration is given as

$$
\begin{cases} y_i(k+1) = y_i(k) = 0\; i \in S_1 \\ y_i(k+1) = \min\limits_{j \in N_i}\{w_{ij} + y_j(k)\}\; i \in S_2 \end{cases}
\tag{7.14}
$$

where $i \in S_1$ represent the PDC candidate and its state remains constant in the whole iteration process. $i \in S_1$ represent the other nodes in the communication network, whose state will update according to the conditions of their neighbors in each iteration. $w_{ij} = \tau_{ij}^{tran} + \tau_{ij}^{prop}$ is the bias term and represents the weights of link ij. N_i represent the neighbors that directly connect with node i.

At the steady state, the state values y_i will converge to y_i^*, shown as [21]

$$\begin{cases} y_i^* = y_i(0) = 0,\ i \in S_1 \\ y_i^* = \lim_{k \to \infty} y_i(k) = \min_{j \in N_i} \left\{ w_{ij} + y_j^* \right\}, i \in S_2 \end{cases} \tag{7.15}$$

Then, the optimal paths between each PMU and the PDC can be found by recursively connecting the parent nodes. For the node i, its parent node P_i can be obtained as

$$\begin{cases} \mathrm{P}_i = \emptyset,\ i \in S_1 \\ \mathrm{P}_i = \arg\min_{j \in N_i} \left\{ w_{ij} + y_j^* \right\}, i \in S_2 \end{cases} \tag{7.16}$$

The information exchange of each node during the iteration relates to their routing tables formulation. There are generally two ways to maintain routing tables, including static routing and dynamic routing. Dynamic routing is known as adaptive routing which changes the routing table according to the change in topology (including line congestion, line breakage, etc.) [27]. Therefore, the proposed algorithm is actually a novel dynamic routing algorithm that helps routers calculate and update the routing tables under different network conditions. The existing dynamic routing algorithm, i.e., OSPF, applies the Dijkstra algorithm to determine the route to each destination [28]. Different from the centralized Dijkstra algorithm, the proposed distributed algorithm only requires the information exchange between the neighboring routers (nodes). The working principle is given in Fig. 7.13.

The details are as follows:

1. For each router, it discovers its neighbors, establish and maintain neighbor relationships;
2. Generate the link state advertisement (LSA), each router will generate its own LSA;
3. Each router exchanges its LSA with its neighbors, assembles the received LSA, and establishes its own LS database (LSDB) (For the Dijkstra algorithm, each router needs to know the LSA of all other routers);
4. Calculate the optimal path according to the proposed algorithm;
5. Load the calculated route into the routing table.

7.3.3 Queuing Delay Check Module

The DBMC algorithm provides an initial solution for path planning between PMUs and the PDC without considering the queuing delay. Then, a queuing delay check

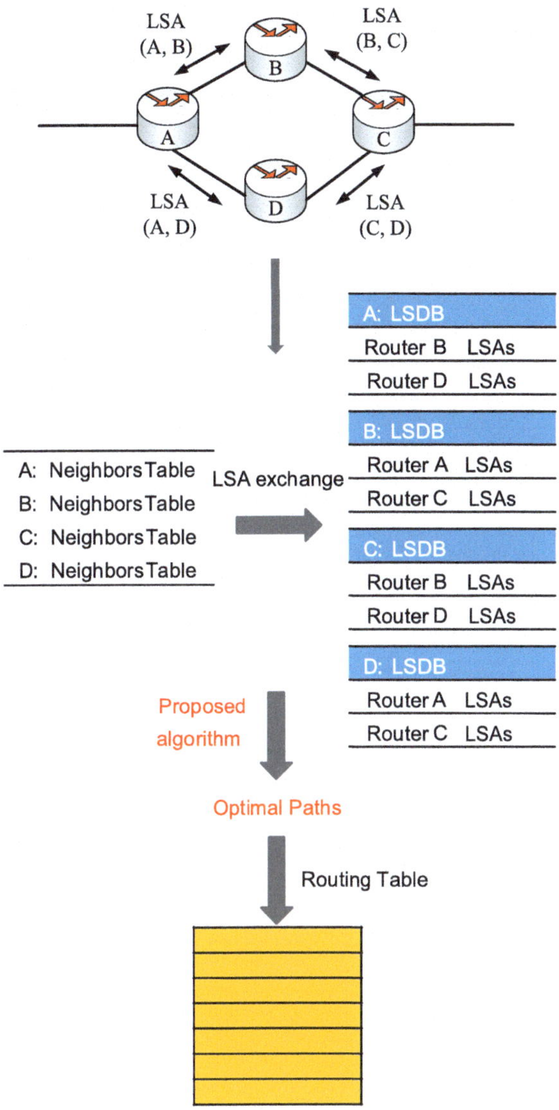

Fig. 7.13 Working principle of the proposed algorithm

module is proposed and applied to check whether the existing paths suffer queuing delay and whether need to be re-selected. The detailed procedure of the optimal path planning considering the queuing delay calculation is given in Fig. 7.14.

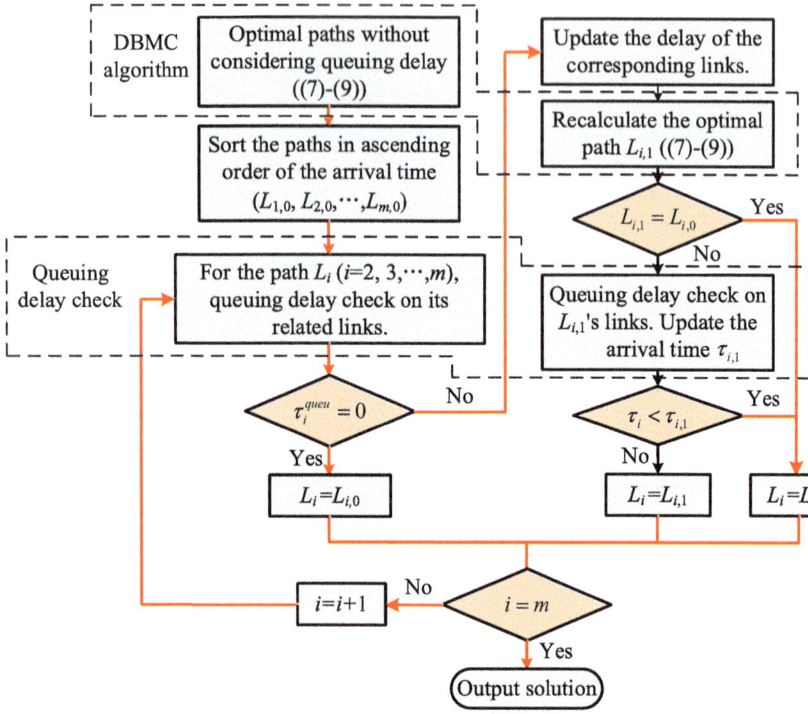

Fig. 7.14 Flow chart of the optimal path planning between PMUs and PDC considering the queuing delay

7.3.4 Optimal PDC Placement

In this section, we first calculate the optimal paths between PMUs and each PDC candidate based on the proposed approach. Then, the PDC candidate with the highest data arrival rate will be selected as the optimal PDC placement, denoted as:

$$\boldsymbol{S}_{PDC} = \arg\max(s_1, s_2 \ldots s_n) \tag{7.17}$$

where s_i $(i = 1, 2, \cdots n)$ is the data arrival rate of PDC candidate i.

2. If there are more than one element in set $\boldsymbol{S}_{\text{PDC}}$, the arrival time of the last data $\boldsymbol{\tau}_{\text{i,last}}$ received by these PDC candidates will be compared further. And the optimal PDC placement is determined as

$$i_{PDC} = \arg\min\left(\tau_{i,\text{last}}\right) \tag{7.18}$$

The links connecting each PMU to i_{PDC} is correspondingly selected as the optimal paths in the communication network.

7.3.5 *Path Replanning Under Cyber-Attacks*

The communication paths might need to be reconstructed when the communication networks suffer cyber-attacks. As mentioned before, the cyber-attacks considered in this chapter include the data transmission delay increase and the communication links being broken. Under these attacks, the network connection relationship and the time delay experienced on this link will be updated first through the router communication. Then, the proposed approach will be executed to obtain the replanned optimal paths between PMUs and the PDC. The details are given as follows.

Algorithm 2: Path replanning under cyber-attacks

Input: Updated communication network topology and link weights
Path Replanning between PMUs and the PDC
Initialize the state variables in the communication network;
Calculate optimal paths between PMUs and the PDC without considering the queuing delay;
Applying the queuing delay module to determine whether the paths should adjust;
Output: Replanned optimal paths connecting each PMU to the PDC

7.4 Case Study

The performance of the proposed approach is tested on the IEEE 39-bus system and 118-bus system. The bandwidth R is set as 1.5 Mbps and 5 Mbps, respectively. The packet size of each PMU data is 52 Bytes [29]. The propagation speed v of the data packet is 2×10^8 m/s [30]. The structure of the communication system is consistent with the corresponding power transmission system, and each node is able to send, receive, and route measurements [14, 15]. The simulation is implemented in MATLAB R2021b on a system with Intel (R) Core (TM) i7 PC, 2.30 GHz processor with 16 GB RAM.

7.4.1 *IEEE 39-Bus System*

Based on (12), nine PMUs are configured at nodes 3, 6, 13, 16, 20, 23, 25, 29, and 39 to make the system observable. Each node is viewed as a PDC candidate. The optimal paths and their ETE delay calculated by the proposed method and the DBMC algorithm are compared. The delays experienced by each path are sorted in descending order, and paths with overlapping links are grouped together. Consider Node 1 as the PDC placement. The corresponding results from the two methods are given in Tables 7.2 and 7.3, respectively. The unit of the ETE delay is millisecond and is abbreviated as *ms* hereinafter.

Table 7.2 Optimal path between each PMU to Node 1 obtained from the DBMC algorithm

PMU	Group	ETE delay (*ms*)	Optimal path
39	0	2.34	39→1
25	1	4.52	25→2→1
3	1	4.54	3→2→1
29	1	9.43	29→26→25→2→1
16	1	11.11	16→17→18→3→2→1
13	1	11.42	13→14→4→3→2→1
6	1	11.72	6→5→4→3→2→1
20	1	15.56	20→19→16→17→18→3→2→1
23	1	15.60	23→24→16→17→18→3→2→1

Table 7.3 Optimal path between each PMU to Node 1 obtained from the proposed algorithm

PMU	Group	ETE delay (*ms*)	Optimal path
39	0	2.34	39→1
25	1	4.52	25→2→1
3	1	6.81 (+2.27)	3→2→1
29	1	9.43	29→26→25→2→1
16	1	11.71	16→17→18→3→2→1
13	1	13.99 (+2.57)	13→14→4→3→2→1
6	1 (Original)	16.27 (+4.55)	6→5→4→3→2→1
	0 (Existing)	11.84	6→7→8→9→39→1
20	1	17.41 (+1.85)	20→19→16→17→18→3→2→1
23	1	20.04 (+4.44)	23→24→16→17→18→3→2→1

It can be concluded that: (1) Some optimal paths obtained by the DBMC algorithm will suffer queuing delay, including PMU 3, 13, 6, 20, and 23, which lead an inexact calculation of ETE delay; (2) The proposed algorithm evaluates whether the paths suffer queuing delay, thus providing a more practical and exact solution of the optimal paths and the ETE delay between PMUs and PDC; (3) If the path suffers queuing delay, the queuing delay check module will help determine if an alternative path should be selected with lower ETE delay (e.g., PMU 6). Therefore, compared with the direct application of the DBMC algorithm, the proposed method can provide a better path solution in the WAMS communication network.

(1) *Optimal Path Planning Between PMUs and PDC and PDC Placement*

The ETE delay distribution under each PDC candidate and the corresponding optimal paths are given in Fig. 7.15.

It can be found that the ETE delay of most PDCs mainly distributed between around 5 ms to 20 ms. The ETE delay of PMU data under some PDC candidates is

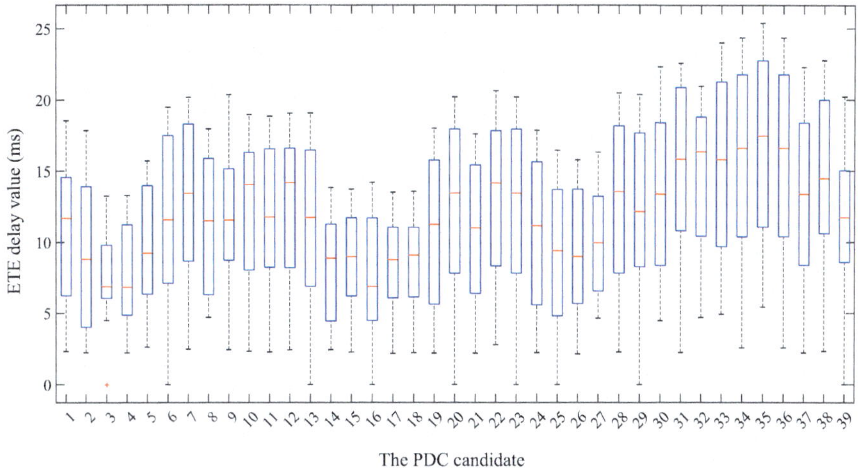

Fig. 7.15 The ETE delay distribution of each PDC candidate

much larger than the average ETE delay values, which reflects that the candidates may not be a good choice for the PDC placement.

The PDC candidates with the maximum data arrival rate under different PDC waiting times are shown in Table 7.4. The PDC waiting time is set to 10 ms, 15 ms, and 20 ms, respectively [29].

Further, the time of the last arrival data of each PDC candidate is compared. Then, node 3 is selected as the optimal PDC placement. The optimal paths connecting PMUs and the PDC 3 are given in Fig. 7.16. The ETE delay experienced and the corresponding node traversed when data transferred by each PMU reaches the PDC 3 are shown in Table 7.5.

(2) *Paths' Reconstruction Under Cyber-Attacks*

According to the above analysis, node 3 is selected as the optimal PDC placement. Each PMU will send its data to the PDC 3 along the path in the normal operation state. However, when the communication links suffer from cyber-attacks, the ETE delay experienced by each data may change, thus affecting the received data completeness of the PDC.

Table 7.4 The PDC candidate with maximum data arrival rate

Waiting time	The PDC candidates with the maximum data arrival rate
10 ms	3 (77.78%)
15 ms	3, 4, 14, 15, 16, 17, 18 (100%)
20 ms	1–6, 8, 10–19, 21, 24–27 (100%)

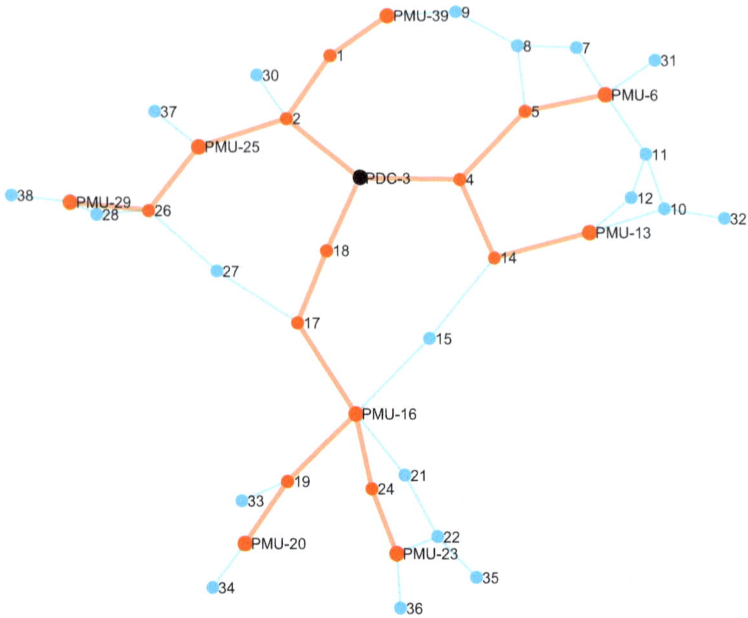

Fig. 7.16 Optimal paths connecting each PMU and the PDC 3

Table 7.5 The ETE delay of each PMU data to the PDC

PMU	ETE delay (*ms*)	Optimal path
3	0	3
25	4.50	25→2→3
16	6.57	16→17→18→3
13	6.87	13→14→4→3
39	6.88	39→1→2→3
6	9.10	6→5→4→3
29	9.40	29→26→25→2→3
20	11.01	20→19→16→17→18→3
23	13.24	23→24→16→17→18→3

In this section, the proposed method is used to reconstruct paths between PMUs and the PDC to address the completeness of received data under cyber-attacks. Two types of cyber-attack are considered, namely:

Case 1: The transmission delays of links (3, 4) and (23, 24) are increased by 40 ms [15];

Case 2: The link (1, 2) and (2, 25) suffer complete breakage.

For the two cases, the PDC waiting time is set to 20 ms. The replanned paths for the two cases in shown in Fig. 7.17.

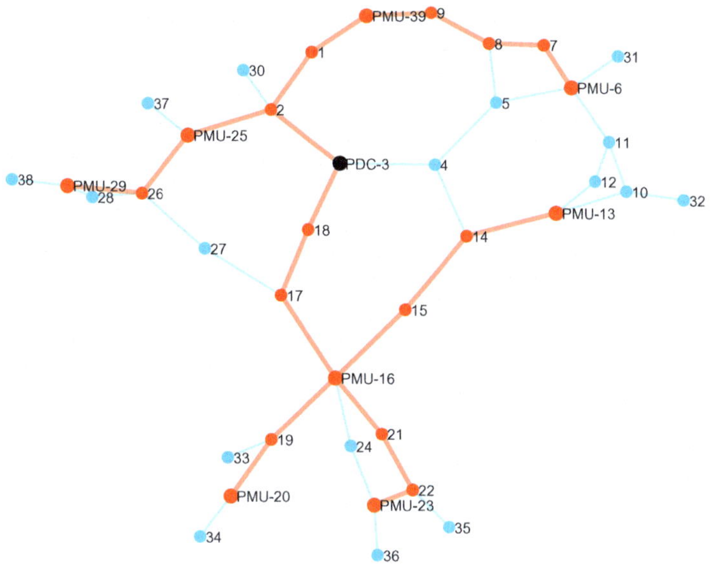

(a) Path reconstruction in Case 1

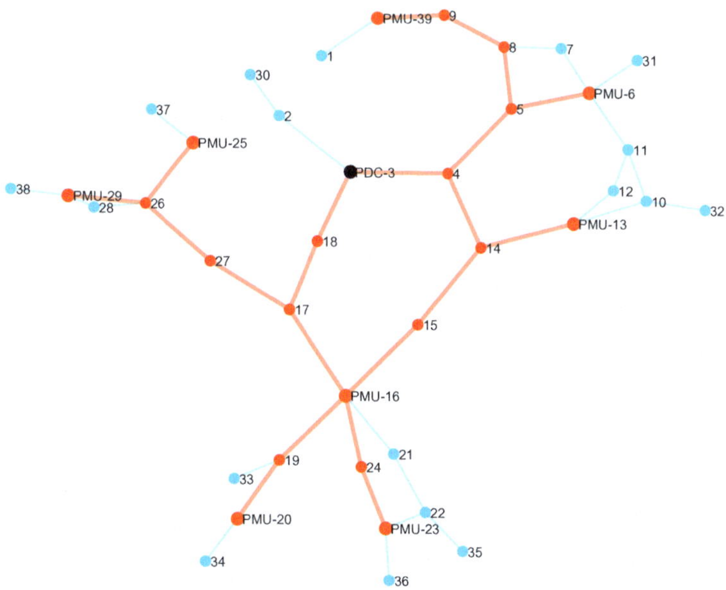

(b) Path reconstruction in Case 2

Fig. 7.17 Path reconstruction under cyber-attacks

It can be found in Fig. 7.17a that the optimal path of PMU 6 adjusts from {6→5→4→3} to {6→7→8→9→39→1→2→3}. And the optimal path of 13 adjusts from {13→14→4→3} to {13→14→15→16→17→18→3}.

Likewise, in case 2, the new optimal path of PMU 25 adjusts from {25→2→3} to {25→26→27→17→18→3}. And the optimal path of 39 adjusts from {39→1→2→3} to {39→9→8→5→4→3}.

Then, the ETE delay experienced by each PMU data under "Normal", "Cyber-attacks", and "Replanning", and their data arrival rates are given in Table 7.6. The PMU data that fails in received by the PDC due to the cyber-attack is denoted as "/".

Due to the cyber-attack, there are two sets of data that cannot be received by PDC 6 in Case 1 and Case 2. The data arrival rate under the two cases decreases to 77.78% and 66.67%, respectively. By paths reconstruction of the proposed approach, the PDC 6 can receive the lost data before and ensures 100% data arrival rate. It proves that this method can address the completeness of received data under cyber-attacks. Further, the computation time for the two cases is 23.94 ms and 27.61 ms, respectively, which validates that the proposed method can meet the online implementation requirements.

(3) *Cyber impact on PMU-based state estimation*

As mentioned before, the ETE delay will influence the completeness of data received by the PDC, thus affecting the performance of the WAMS applications. In this section, the cyber impact of the ETE delay toward the state estimation is detailed.

To analyze the cyber impact on the state estimation, the performances of state estimation under different PDC placements and the corresponding paths are compared. The performance index is selected as the maximum average estimated errors, expressed as [29, 31]

$$P(\mathbf{x}) = \frac{1}{T}\sum_{i=1}^{T} \max\left|(\tilde{x}_i^t - x_i^t)/x_i^t\right| \tag{7.19}$$

Table 7.6 The ETE delay of each PMU data and the data arrival rated under three states

PMU	Normal (*ms*)	Cyber-attacks (*ms*)		Replanning (*ms*)	
		Case 1	Case 2	Case 1	Case 2
3	0	0	0	0	0
25	4.50	4.50	/	4.50	13.27
16	6.57	6.57	6.57	6.57	6.78
13	6.87	/	6.87	6.88	9.01
39	6.88	6.88	/	9.40	11.25
6	9.10	/	9.10	11.01	6.78
29	9.40	9.40	9.40	13.45	15.50
20	11.01	11.01	11.01	15.68	11.03
23	13.24	13.24	13.24	16.38	14.40
Arrival rate	100%	77.78%	66.67%	100%	100%

Table 7.7 State estimation results under different PDC placement and their corresponding paths

PDC	Data arrival rate (%)	Power System observability	$P(\boldsymbol{u})$	$P(\boldsymbol{\theta})$
3	100	Observable	1.35×10^{-2}	8.62×10^{-2}
31	66.67	Unobservable	/	/
35	55.56	Unobservable	/	/

where $\tilde{x}_i^t$ and x_i^t are the estimated and true values of node i in the *kth* experiment. The measurements transmitted by PMUs include the node voltage phasor and line current phasor. T is the total number of the Monto-Carlo trials, which take 1000 in this chapter.

Then, the performance of state estimation under three PDC placements and their corresponding paths is compared, i.e., PDC 3 (the optimal PDC placement), PDC 31, and PDC 35. The data arrival rate and the maximum average estimated errors under three cases are given in Table 7.7. The PDC waiting time is set to 20 ms.

The observability of the power system is determined by the rank of the Jacobian matrix $\boldsymbol{H}$ of the measurement data received by the PDC [29]. If $\boldsymbol{H}$ is full rank, the power system is observable and the state estimation can be executed. From Table 7.7, it can be found that the power system is observable under PDC 3 and its path construction between PMUs. Then, an accurate estimate of the system state is obtained based on the transmitted measurement date from PMUs. As for PDC 31 and 35, they failed to receive measurements from some PMUs, thus leading the power system to be unobservable and the state estimation cannot be executed.

In conclusion, the PDC placement and its corresponding paths between PMUs and the PDC will influence the ETE delay suffered by each PMU data and the PDC's data arrival rate. The completeness of data received by the PDC determines the systems' observability and state estimation accuracy. The proposed method can improve data completeness obtained by the PDC, thereby enhancing the state estimation performance.

7.4.2 *IEEE 118-Bus System*

There are 29 nodes installed with PMUs to guarantee the system's observability, which is given in Table 7.8.

For each PDC candidate, their optimal paths connecting with PMUs are calculated by the proposed approach, and the ETE delay experienced by each PMU is recorded.

Table 7.8 PMUs configuration of IEEE 118-bus system

PMUs configuration
3, 9, 11, 12, 17, 21, 25, 29, 34, 37, 40, 45, 49, 52, 56, 62, 65, 72, 75, 77, 80, 85, 87, 91, 94, 101, 105, 110, 114

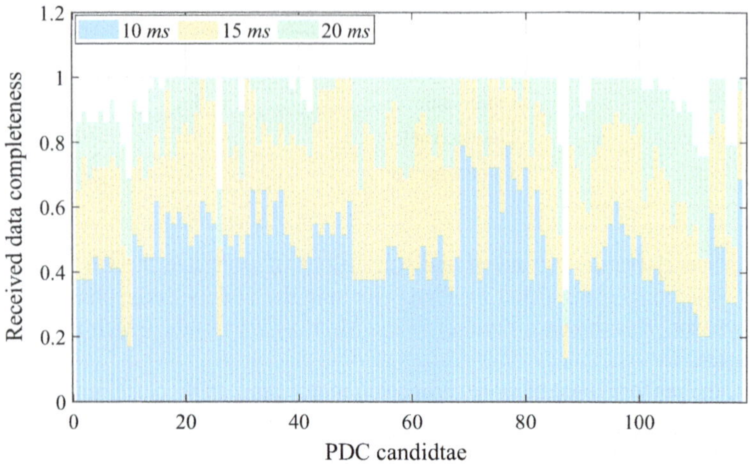

Fig. 7.18 The ETE delay distribution of each PDC candidate under three waiting time settings

Then, the received data completeness of each PDC candidate under three PDC waiting times is compared and given in Fig. 7.18.

By comparison, the optimal PDC placement is determined to be set at node 69. The optimal paths between PDC 69 and the 29 PMUs are shown in Fig. 7.19. To visually show the ETE delay experienced by each PMU data when it arrives at the PDC, and to determine whether there is a queuing delay, the ETE delay suffered by each PMU data is given in Fig. 7.20. The paths are divided into six groups according to the overlapping links they pass through.

In each group (i.e., subgraph), the horizontal axis represents the PMU number. The bar of each PMU is divided into several blocks, and the number of blocks represents the number of links the PMU data passes through. The block with the same color represents that the PMUs pass through the same links, whereas the white blocks refer to only one PMU data passing through those links. The length of the block represents the ETE delay experienced on this link. It can be found that the time that passes through the same link for PMUs may be different due to the queuing delay. Take Fig. 7.19d as an example, the ETE delay experienced by PMU 72, 25, 62, 114, 21, 17, 29, and 11 through the link "24–70" varies, with PMU 21 suffering the most extended ETE delay on that link.

Then, the performance of the proposed method in replanning paths is tested when multiple original paths are destroyed due to the cyber-attack. Here, there are three cases, that is, 2 paths, 4 paths, and 6 paths suffer breakage. In each case, 100 Monte Carlo tests from C_m^n situations were carried out. The average data arrival rate after the path replanning and its running time were calculated under PDC waiting times of 15 ms and 20 ms. The results are shown in Table 7.9.

As can be seen from Table 7.10, the reconstruction path provided by this method can still guarantee the completeness of the data received by PDC when the original

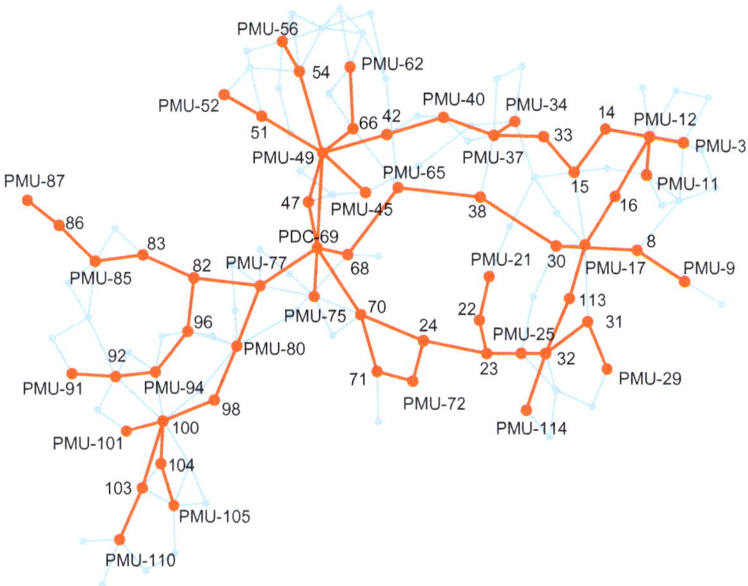

Fig. 7.19 Optimal paths connecting PMUs and the PDC 69

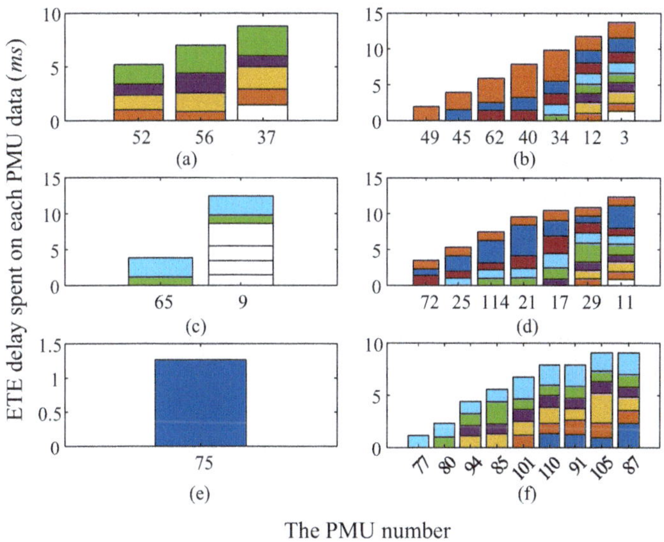

Fig. 7.20 The ETE delay experienced by each PMU before reaching the PDC 69

Table 7.9 The performance of the method for path replanning under cyber-attacks

Number of damaged paths	Average data arrival rate		Average running time (*ms*)
	15 ms (%)	20 ms (%)	
2	100	100	37.13
4	97.38	99.66	31.60
6	93.66	98.69	34.28

damage path is few. As the number of damaged paths increases, the number of alternative paths in the communication network decreases. Therefore, the completeness of the corresponding received data is reduced. But they are still at a relatively high level. Meanwhile, the average computation time shows that the proposed method can meet the real-time requirements in the large test system.

7.4.3 Comparison with the Existing Methods

This section compares the performance of the proposed method with the existing methods [12, 14]. The optimal PDC placement and its data arrival rate are considered as the performance index. The results are shown in Table 7.10. The PDC waiting time is set to 10 ms.

It can be seen that the data arrival rate under the selected PDC placement and the planned paths from [12] are not good in the two test systems, since it only considers the physical length of the links but neglects the ETE delay experienced on the link. The method in [14] takes the transmission and propagation delay into account, while missing the queuing delay. The data arrival rate of [14] reaches 100% in the small test system. However, as the size of the system and the amount of data transmitted increase, the queuing delay is exacerbated, and the data arrival rate decreases obviously in the large test system. The method proposed in this chapter achieves a 100% data arrival rate in both test systems. It is verified that this method can effectively solve the data completeness problem received by the PDC.

Table 7.10 Results comparison of the proposed method and the existing methods

Method	Optimal PDC placement		Data arrival rate	
	IEEE-39	IEEE-118	IEEE-39 (%)	IEEE-118 (%)
[12]	13	22	66.67	79.31
[14]	18	68	100	75.86
Proposed method	3	69	100	100

7.5 Conclusion

To minimize the ETE delay and maximize the completeness of the PMU data received by the PDC in WAMS, a distributed DBMC-based method is proposed to optimize the communication network. A queuing delay detection module is considered to check the queuing delay in optimizing the paths between PMUs and the PDC. At the same time, the influence of the completeness of data received by PDC on the application of state estimation is discussed in detail. Numerical tests verify that the proposed DBMC-based method can provide an effective solution and meet the real-time requirements of communication links under normal conditions and cyber-attacks. Future work will focus on layout optimization methods for multiple PDCs in WAMS communication networks and the design and implementation of test benches.

References

1. S. Das, B.K. Panigrahi, A PMU-based data-driven approach for enhancing situational awareness in building a resilient power systems. IEEE Trans. Ind. Inform. **18**(7), 4773–4784 (2022)
2. G. Ravikumar, D. Ameme, S. Misra, S. Brahma, R. Tourani, ICASM: an information-centric network architecture for wide area measurement systems. IEEE Trans. Smart Grid **11**(4), 3418–3427 (2020)
3. B. Naduvathuparambil, M.C. Valenti, A. Feliachi, Communication delays in wide area measurement systems, *Proceedings of the Thirty-Fourth Southeastern Symposium on System Theory* (Cat. No.02EX540), pp. 118–122 (2002)
4. J. Zhao et al., Roles of dynamic state estimation in power system modeling, monitoring and operation. IEEE Trans. Power Syst. **36**(3), 2462–2472 (2021)
5. M.E.C. Bento, A hybrid particle swarm optimization algorithm for the wide-area damping control design. IEEE Trans. Ind. Inform. **18**(1), 592–599 (2022)
6. M. Rezaee, M.H. Yaghmaee Moghaddam, SDN-based quality of service networking for wide area measurement system, IEEE Trans. Ind. Inform. **16**(5), 3018–3028 (2020)
7. H.H. Alhelou, P. Cuffe, A dynamic-state-estimator-based tolerance control method against cyberattack and erroneous measured data for power systems. IEEE Trans. Ind. Inform. **18**(7), 4990–4999 (2022)
8. A.S. Musleh, G. Chen, Z.Y. Dong, A survey on the detection algorithms for false data injection attacks in smart grids. IEEE Trans. Smart Grid **11**(3), 2218–2234 (2020)
9. D. Bertsimas, M. Sim, Robust discrete optimization and network flows. Math. Program. **98**(1), 49–71 (2003)
10. A. Kasperskia, P. Zielinski, The robust shortest path problem in seriesparallel multidigraphs with interval data. Oper. Res. Lett. **34**(1), 69–76 (2006)
11. M. Shahraeini, M.S. Ghazizadeh, M.H. Javidi, Co-optimal placement of measurement devices and their related communication infrastructure in wide area measurement systems. IEEE Trans. Smart Grid **3**(2), 684–691 (2012)
12. M. Beg Mohammadi, R. Hooshmand, F. Haghighatdar Fesharaki, A new approach for optimal placement of PMUs and their required communication infrastructure in order to minimize the cost of the WAMS, IEEE Trans. Smart Grid **7**(1), 84–93 (2016)
13. A. Ghasemkhani, H. Monsef, A. Rahimi-Kian, A. Anvari-Moghaddam, Optimal design of a wide area measurement system for improvement of power network monitoring using a Dynamic multiobjective shortest path algorithm. IEEE Syst. J. **11**(4), 2303–2314 (2017)

14. B. Appasani, D.K. Mohanta, Co-optimal placement of PMUs and their communication infrastructure for minimization of propagation delay in the WAMS. IEEE Trans. Ind. Inform. **14**(5), 2120–2132 (2018)
15. R. Kateb, et al., Enhancing WAMS communication network against delay attacks, IEEE Trans. Smart Grid **10**(3), 2738–2751 (2019)
16. H. Lin, Yi Deng, S. Shukla, J. Thorp, L. Mili, Cyber security impacts on all-PMU state estimator-a case study on co-simulation platform GECO, *2012 IEEE Third International Conference on Smart Grid Communications (SmartGridComm)*, pp. 587–592 (2012)
17. R. Kateb, M.H.K. Tushar, C. Assi, M. Debbabi, Optimal tree construction model for cyber-attacks to wide area measurement systems. IEEE Trans. Smart Grid **9**(1), 25–34 (2018)
18. H.M. Khalid, J.C.-H. Peng, A bayesian algorithm to enhance the resilience of wams applications against cyber attacks, IEEE Trans. Smart Grid **7**(4), 2026–2037 (2016)
19. B. Moussa, A. Al-Barakati, M. Kassouf, M. Debbabi, C. Assi, Exploiting the vulnerability of relative data alignment in phasor data concentrators to time synchronization attacks. IEEE Trans. Smart Grid **11**(3), 2541–2551 (2020)
20. Y. Cui, F. Bai, Y. Liu, P.L. Fuhr, M.E. Morales-Rodríguez, Spatio-temporal characterization of synchrophasor data against spoofing attacks in smart grids. IEEE Trans. Smart Grid **10**(5), 5807–5818 (2019)
21. Y. Zhang, S. Li, Distributed biased min-consensus with applications to shortest path planning. IEEE Trans. Automat. Contr. **62**(10), 5429–5436 (2017)
22. Y. Xu, H. Sun, W. Gu, A novel discounted min-consensus algorithm for optimal electrical power trading in grid-connected DC microgrids. IEEE Trans. Ind. Electron. **66**(11), 8474–8484 (2019)
23. P. Yao, R. Zhao, Q. Zhu, A hierarchical architecture using biased min-consensus for USV path planning. IEEE Trans. Veh. Technol. **69**(9), 9518–9527 (2020)
24. A. Lang, Y. Wang, C. Feng, E. Stai, G. Hug, Data aggregation point placement for smart meters in the smart grid. IEEE Trans. Smart Grid **13**(1), 541–554 (2022)
25. M. Göl, A. Abur, LAV based robust state estimation for systems measured by PMUs. IEEE Trans. Smart Grid **5**(4), 1808–1814 (2014)
26. D. Dua, S. Dambhare, R. K. Gajbhiye, S. A. Soman, Optimal multistage scheduling of PMU placement: an ILP approach. IEEE Trans. Power Deliv. **23**(4), 1812-1820 (2008)
27. T. Mak, P.Y.K. Cheung, K.-P. Lam, W. Luk, Adaptive routing in network-on-chips using a dynamic-programming network. IEEE Trans. Industr. Electron. **58**(8), 3701–3716 (2011)
28. R.A. Guerin, A. Orda, D. Williams, QoS routing mechanisms and OSPF extensions, GLOBECOM 97. IEEE Global Telecommunications Conference. Conf. Rec. **3**, 1903–1908 (1997)
29. M. Chenine, L. Nordström, Investigation of communication delays and data incompleteness in multi-PMU wide area monitoring and control systems, *2009 International Conference on Electric Power and Energy Conversion Systems, (EPECS)*, pp. 1–6 (2009)
30. X. Liu, S. Zhang, X. Zeng, L. Yao, Y. Ding, C. Deng, Evaluating the network communication delay with WAMS for multi-energy complementary systems. CSEE J. Power Energy Syst. **6**(2), 402–409 (2020)
31. Y. Chen, Y. Yao, Y. Zhang, A robust state estimation method based on SOCP for integrated electricity-heat system. IEEE Trans. Smart Grid **12**(1), 810–820 (2021)

Chapter 8
Optimal Coordination of District-Scale Multi-Energy Systems

Michael Taylor, Jinglin Liu, Ognjen Marjanovic, and Alessandra Parisio

Abstract A major challenge in the transition to a net-zero energy system revolves around the decarbonization of energy use for heating, cooling and transport via electrification, whilst simultaneously ensuring the security of a power system with high penetration of renewable energy generation. One possible way to address this challenge is to adopt a multi-energy systems approach, in which traditionally separate energy systems for the delivery of electricity, gas, heating and cooling are co-optimized as an integrated entity. A major benefit of this approach is that flexible distributed energy resources in non-electrical systems can be utilized to support the power grid. This chapter presents a novel multi-energy system optimization modeling framework, capable of rapidly formulating large-scale optimization problems. These problems can be readily integrated within model predictive control (MPC) schemes, providing a method for online energy management of a continuously evolving system. Such schemes can optimally manage distributed energy resources at a community or *district* scale, ensuring that all energy demands are met, networks are operated within acceptable safety limits and cost savings are delivered to both customers and network operators. Given the potentially large size of the resulting control problem, a multi-agent control architecture and associated coordination algorithms are also presented. These ensure that near-optimal, feasible control actions can be determined within timescales that are suitable for online energy management. In an exemplary case study, considering a 15 min sampling interval for a district comprising 84 buildings and multiple energy supply networks, a maximum computation time of approximately 55 min for a single controller is reduced to just over 1 s using the novel multi-agent MPC scheme, demonstrating the substantial benefit of the proposed approach.

M. Taylor · J. Liu · O. Marjanovic · A. Parisio (✉)
Department of Electrical and Electronic Engineering, School of Engineering, University of Manchester, Oxford Road, Manchester M13 9PL, UK
e-mail: alessandra.parisio@manchester.ac.uk

J. Hu (ed.), *Community Energy and Microgrids*, Green Energy and Technology,
https://doi.org/10.1007/978-981-97-6297-2_8

8.1 Introduction

The significance of community energy within the context of the net-zero transition cannot be overstated. Communities equipped with their own energy assets play a pivotal role in fostering energy-efficient practices, optimizing self-consumption, and mitigating the strain on the primary grid. This role becomes practically critical when considering the formidable challenges faced by power system operators in ensuring grid stability.

Power systems must operate within narrow frequency and voltage bounds to maintain stability, since any deviations in these metrics can lead to serious and far-reaching consequences, such as load shedding or blackout. Historically, power system operators have relied on substantial, controllable generators to address these challenges. However, with continued phase out of fossil fuel-driven generators, this controllable capacity will inevitably dwindle.

Traditionally, system operators have had limited control over the system load and would rely on the generation capacity of large, nuclear- or fossil fuel-driven synchronous generators (operating at nominal frequency) to balance the system. As coal- and gas-fired power plants are phased out as part of the transition to net-zero, system operators will no longer have this large amount of controllable capacity at their disposal. Instead, the supply mix will become dominated by unscheduled wind and solar, thereby reducing supply-side flexibility and complicating the task of adjusting generation to meet the demand. Furthermore, these renewable sources lack the inertia possessed by synchronized generators, which is crucial for absorbing short-term fluctuations in frequency. Additionally, non-synchronous generators, such as wind turbines and solar photovoltaics (PV), exhibit limited voltage regulation capabilities, further exacerbating voltage stability issues, particularly when coupled with weak connections to the main grid [1].

To confront these pressing issues, the concept of demand-side flexibility, with a particular focus on demand-side response, is gaining traction. Demand-side response encompasses the management of electricity consumption on the customer's side of the meter, including load shedding, demand shifting, and on-site generation utilization. A variety of distributed energy sources, such as flexible loads, energy storage, and distributed generation, can contribute to bolstering demand-side flexibility.

In particular, electric vehicles equipped with smart charging capabilities and buildings heated or cooled using electric heat pumps inherently possess energy storage potential [2, 3], rendering them invaluable for demand-side response. Moreover, ultra-low-temperature district heating and cooling (ULTDHC) networks offer the prospect of thermal energy storage [4]. The flexibility derived from these distributed energy sources holds the potential to yield substantial savings in power system infrastructure and operational costs. Flexibility managed across heat, transport, industry and power sectors in Great Britain was recently estimated to potentially save £9.6–16.7 billion/year assuming a net-zero scenario in 2050 [5].

Regrettably, these benefits are yet to be fully acknowledged in European energy retail markets, in which the time-varying nature of supply is masked by the prevailing use of flat electricity price tariffs. To unlock the potential of flexibility, both producers and consumers (who may also generate energy) require incentives, such as the implementation of dynamic rates in the retail market [6]. While demand response services can be traded in wholesale markets, smaller producers often encounter barriers to participation due to their scale. This has led to the emergence of *aggregators*, serving as intermediaries that facilitate the coordination of multiple distributed energy sources [7].

When considering commercial aggregation across a community or district, the advantages become apparent, including the scale and diversity of resources, and the capacity to ensure coordinated responses to network challenges, by efficiently managing resources across electricity, natural gas, heating, and cooling networks [6].

Nevertheless, the optimization of multi-energy districts presents formidable technical challenges [8], stemming from the combination of interconnected systems, non-linear equations, and the complexity of non-convex modeling of battery storage systems or multi-mode energy conversion devices [9, 10]. Addressing these complexities is paramount, and one approach may involve decomposing the global optimization problem into smaller sub-problems, though this approach is not without its own set of challenges, especially when dealing with non-convex formulations.

Efficient modeling representations and optimization methodologies hold the key to realizing the vision of multi-energy districts, promising substantial cost savings, enhanced reliability and a reduction in emissions. Consequently, addressing these formidable challenges remains at the forefront of ongoing research in this field.

In [11] and [12], a new modeling method for pressure drop analysis of ULTDHC networks is introduced and it is shown how the model can be integrated with AC current analysis and distributed energy management models in a multi-agent MPC scenario. This chapter presents some outcomes from these works, which are intended to address the technical challenges outlined above.

8.2 Literature Review

This section discusses some of the approaches found in the literature for modeling multi-energy systems, together with an introduction to Ultra-Low-Temperature District Heating Networks (ULTDHC), which cannot be modeled with sufficient accuracy using the approaches available. Following this, examples of online multi-energy system coordination from the literature are reviewed in Sect. 8.2.1.

8.2.1 Multi-Energy System Modeling

For optimization problems in multi-energy system management, equality and inequality constraints are used to describe the physical or operational limits of undetermined system variables, as well as the physical relationships between them; the constraints provide a representative mathematical model of the system. A convenient level of detail at which to model resources in large, complex energy systems is to focus on steady-state energy flows. These are the quantities for which there may be limited capacity at various points within a system and which strongly correlate to overall operating costs, since commodities such as electricity and gas are priced per unit of energy. Models therefore typically describe the production, consumption, conversion, storage and loss of energy during operation. However, these processes in real, dynamic systems are inherently non-linear and often involve discrete states of operation, i.e. they are *hybrid dynamical* systems. Therefore, the main challenge for multi-energy system modeling is to formulate constraints that represent real systems relatively accurately, whilst also ensuring that the problem can be solved efficiently and remain tractable.

8.2.1.1 Mixed Logical Dynamical Systems

In most of the multi-energy system examples found in the literature, mixed-integer linear programming (MILP) is used to model and optimize the energy systems of interest [9, 10, 13–29]. This stems from the fact that non-linear system behaviour can be approximated by converting logical statements into mixed-integer linear equality and inequality constraints. Characterization of such hybrid systems has been formalized using a compact representation known as the mixed logic dynamics (MLD) framework [30]. The method introduces additional continuous and discrete auxiliary variables z and δ into the well-known discrete-time state-space representation of the system. However, the conversion of logical constraints into inequalities in the MLD framework may be unintuitive and can hinder the generation of MLD models [10]. One option is to use the HYSDEL parser software [31], yet the reliance on another layer of software with rigid functionality is restrictive when designing an energy management scheme. This highlights another challenge in modeling multi-energy systems, namely, the development of frameworks to rapidly generate the optimization constraints required for potentially large systems.

8.2.1.2 Modular Multi-Energy Management Framework

The modeling framework presented in [10, 30] presents arguably the most computationally efficient and straightforward approach for modeling hybrid systems available in the literature. The framework consists of individual sub-system modules using an MLD representation.

Since the approach is intended specifically for multi-energy systems, the difficulty in generating constraints is avoided by having a set of pre-defined modules with which to model distributed energy resources. Modules may be interconnected using any configuration and a general cost function is provided, which allows a user to encourage desirable operational behaviour, e.g. to minimize the rate of change of a given decision variable. The set of pre-defined modules includes energy conversion devices, controllable energy sources and sinks, predictable (uncontrollable) energy sources and sinks, simple energy balances and storage devices. Hence, a key disadvantage with this, and all the other existing frameworks, is the lack of support for detailed, computationally efficient modeling of physical network flows in conjunction with distributed energy resources. The framework is therefore limited to applications involving relatively small systems, for which energy losses, voltage drop and pressure drop may be negligible, or applications where network operating conditions might justifiably be ignored.

8.2.1.3 Network Modeling and Analysis

Network conditions do not have to be ignored in practice and modeling is typically carried out using:

- alternating current (AC) power flow models of voltage, current, active and reactive power in electricity networks [32];
- gas flow models of mass and pressure in gas networks [33] and
- thermal models relating mass, pressure and temperature in district thermal networks [34].

Each of these flow models, which are herein collectively referred to as energy flow models, involves highly non-linear, non-convex relationships. Power flow models include the bus injection model and the less common branch flow model, which are used to represent steady-state, balanced, single-phase networks [32]. The branch flow model is of particular interest for optimization because it can be represented by a convex second-order cone programming (SOCP) form for which solutions are guaranteed to be exact, i.e. feasible for the original problem, under mild conditions [35]. The gas flow model, typically based on the popular steady-state Weymouth equation to determine pressure drop [33], can also be relaxed to a mixed-integer conic form which has been shown to be exact for radial gas networks that have one fixed pressure node [36]. Finally, district thermal networks require both a hydraulic model and a thermal model, each based on steady-state conservation of mass and energy [34, 37]. Both of these models are non-convex and attempts to provide a convex formulation have either involved a conic relaxation of the hydraulic model and linearization of the thermal model, by ignoring thermal losses [38], or linearization of both models by assuming fixed mass flow rates [39–42]. However, district heating networks with constant mass flow rates have been shown to exhibit worse performance than those employing variable flows [34] and the assumption of negligible thermal losses to the surroundings is only valid for networks operating close to ambient temperatures, known as ultra-low-temperature district heating (ULTDH) networks [43]. District heating networks have undergone a technological evolution spanning 140 years, with a gradual reduction in operating temperatures due to changes in socio-economic factors [44]. The first generation of such networks used pressurized steam as the heating medium with temperatures below 200 °C, whereas by the fourth generation, operating temperatures had decreased to 50–60 °C, enabling the integration of renewable energy sources for both electricity and heat. Following the global trend towards decarbonizing energy sources, further lowering of operating temperatures as well as increasing efficiency and substituting heat sources have become a prevailing trend for district heating networks. ULTDH represents a specific configuration of 4GDH, operating at temperatures below 50 °C. This implies that the medium cannot be used for direct heating purposes, and each substation connecting the network to a given building must be equipped with a water-source heat pump to heat the hot water to the required temperature. The substations can also operate in reverse to provide cooling for buildings, known as Ultra-Low Temperature District Heating and Cooling (ULTDHC). When used for heating, it draws warm water from the hot pipe, extracts heat using WSHP (water-source heat pump) and discharges the cooling water into the cold pipe, and vice versa when used for cooling. Thus both heating and cooling can be supplied simultaneously by a single two-pipe network, reducing installation cost and complexity. This is a unique advantage not shared with other, higher temperature 4GDH configurations, which deliver heat from heat producers to end-users by a series of uni-directional supply and return pipes. When using such uni-directional configurations, the provision of cooling would have to be delivered by a separate cooling network [43, 44].

The ULTDHC network can either have a centralized pumping station or decentralized pumps at each substation (Fig. 8.1), allowing users to switch between heating and cooling demands at different times (bi-directional mass and energy flow), enabling them to become heat prosumers [45]. The heating and cooling demands of buildings are dynamically balanced by the district network in real time. However, seasonal variations will result in heating or cooling demand dominating at different times. Therefore, some form of large-scale thermal storage facility, such as aquifers or borehole thermal energy storage, is needed to manage seasonal imbalances. If there is an imbalance in cumulative heating and cooling demands over a year, external heating or cooling supply may also be required.

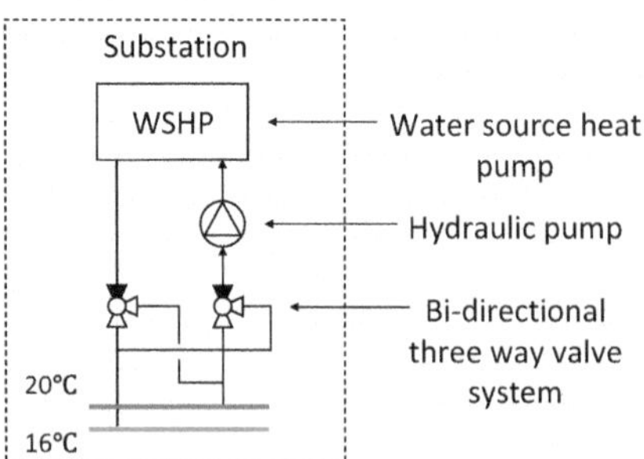

Fig. 8.1 Schematic of an example ULTDHC substation, capable of bi-directional operation using a reversible WSHP, based on [45]

8.2.2 Multi-Energy System Management

Approaches that prioritize device modeling are often associated with online energy management control of a multi-energy system, e.g. [10, 15, 21, 22, 29, 46].

The corresponding optimization problem is incorporated into a model predictive control (MPC) scheme and then, due to the inclusion of MLD system constraints [30], solved using mixed-integer linear programming (MILP). MLD constraints can be used to approximate non-linear behaviour or represent discontinuous device characteristics whilst being relatively easy to solve using off-the-shelf MILP solvers. Many optimal scheduling techniques can be repurposed for online control by implementing an MPC or *receding horizon* strategy, which possesses the following common features [47]:

- explicit use of a model to predict system outputs over a time horizon;
- calculation of a control signal by minimizing an objective function and
- receding horizon updates, in which the first instance of the control signal is applied before the time horizon is displaced one time step towards the future and a new control signal is recalculated.

An MPC scheme is a multi-variable strategy integrating online optimization of a prediction model with a feedback mechanism which can partly compensate for forecast and modeling errors, as well as the unanticipated and unmeasured disturbances, whilst respecting constraints imposed on various system variables [47–50]. All of these features make MPC ideal for energy management applications which generally involve a large number of variables, dynamic models, operational constraints and a degree of uncertainty in forecast variables [9]. However, computation time is a critical factor when deploying optimal online control. If a feasible, optimal (or at least near-optimal) control action cannot be provided within a limited period after a system state measurement or estimate is obtained, then this action will be based on out-of-date information and may no longer be optimal or even feasible. Furthermore, if the worst-case computation time is longer than the sampling interval length, an even more problematic situation may arise in which no control action is provided at all. As a result, there is typically a compromise made regarding either the modeling accuracy, system scale, prediction horizon length or sampling frequency when designing online energy management control schemes (see Table 8.1).

For example, an MPC scheme for a microgrid with a combined cooling, heat and power system is presented in [16]. In this application, MPC aims to minimize energy production costs by solving an MILP problem every 15 min, using short-term error prediction and static optimization-based feedback correction every 5 min. The computation times of the rolling optimization and feedback correction are sufficiently small, averaging 2.2 and 1.1 s, respectively. However, the prediction horizon length of 4 h is relatively short and the size of the system is also relatively small, with only two storage systems and four controllable energy conversion devices, therefore the resulting MILP problem is small and easily solved.

Table 8.1 Overview of model predictive control studies reviewed in this section

Ref.	Prediction horizon	Sampling interval	Connected systems	Networks modeled	Uncertainty handling	Method	Control structure
[16]	4 h	15 min	1	n/a	Det.	MILP	SC
[21]	24 h	30 min	1	n/a	Det.	MIQP	SC
[10]	24 h	15 min	1	n/a	Det.	MILP	SC
[28]	3–7 days	7.5–15 mins	1–12	HTDH & DC	Det.	MILP	SC
[51]	24 h	1 h	6	None	ANN	NLP	SC
[52]	150 s	30 s	1	n/a	SO	QP	SC
[53]	10 s	1 s	1	n/a	SO	QP	Hierarchical
[19]	24 h	30 min	1000	None	SO	MILP + RBC	Hierarchical
[22]	3–24 h	0.5–1 h	1	n/a	Det.	MILP + RBC → LP	SC
[20]	24 h	10 min	1	n/a	Det.	LP	SC
[54]	24 h	15 min	3690	Power	Det.	MILP	Multi-agent
[39]	24 h	1 h	4	Power, gas & HTDH	Det.	MISOCP	Multi-agent
[15]	24 h	10 min	5–15	None	Det.	MILP	Multi-agent

ANN = Artificial Neural Network; DC = District Cooling; HTDH = High-Temperature District Heating; RBC = Rule-Based Control; SO = Stochastic Optimization; Det. = Deterministic; SC = Single controller

The scale and complexity of the system under consideration is increased in [21], which comprises three university campus buildings. One of the buildings can consume heat flexibly, whilst another can consume electricity flexibly, providing additional demand response capacity. The controller optimizes an economic objective to minimize costs and changes to flexible loads, resulting in an MIQP problem, for a prediction horizon of 24 h which is discretized into 48 half hourly periods. Although the reported computation time for the control problem of 32 s is appropriate given the sampling frequency, if the system scale were increased to the district scale then the problem could become intractable.

A similarly sized system was studied in [10] which utilized more detailed device modeling by virtue of the adopted MLD modeling formulation, although it did not include flexible loads. A very similar MPC scheme to that of [21] was presented, solving a single MILP problem every 15 min with a prediction horizon of 48 h. The computation time was surprisingly fast at approximately 1.3 s, which may be attributed to the use of a high-performance server to run the computation and possibly fewer binary variables in comparison to [21]. However, no networks were modeled in detail in either study and the problem of scalability for MILP problems remains.

This problem of scalability was also demonstrated in [28], wherein an MILP problem is designed to manage the thermal comfort of up to 12 buildings, interconnected by a high-temperature district heating network and separate cooling network. Due to the slow thermal dynamics considered, a prediction horizon ranging from 3 days

Table 8.2 Computational time for case studies of varying problem size reported in [28]

No. of buildings	Prediction horizon (days)	Sampling interval	Computation time
1	7	15 min	<5 min
12	7	7.5 min	5 h
12	3	7.5 min	20 min

to 1 week is considered, for which the resulting computational times are given in Table 8.2. Clearly this scheme would require amendment if it were to be used in a receding horizon implementation, as is suggested by the authors, since the computation time for 12 buildings exceeds the sampling interval. In order to render this scheme implementable, the network model is simplified by considering fixed mass flows and uniform supply temperatures.

The approach proposed in [51] avoids MILP optimization altogether, favouring a combination of non-linear modeling and an artificial neural network. The reason given for avoiding MILP is that it cannot accommodate the non-linear functions used to represent devices operating at part load, although these can be approximated using MILP, as discussed in [10]. The artificial neural network is used to predict the indoor temperature of an office building in response to a controllable set-point temperature. This office temperature prediction model and non-linear models for a heat pump, gas boiler and CHP are incorporated into a genetic algorithm optimization solver, implemented every hour with a 24 h prediction horizon. Despite having relatively few controllable resources and only four fixed heating loads from other buildings, the non-linear optimization takes 143 s to compute a control action, which is slow for a system of this size.

At the other end of the spectrum of control problem complexity, stochastic optimization schemes for a hydrogen-based microgrid in [52] make use of models, which have been linearized around an operating point. The increased computational burden encountered with scenario-based stochastic MPC methods is offset by a very short prediction horizon of 150 s and very small system size, enabling a sampling interval of 30 s.

The aforementioned examples highlight the compromises made in centralized single-controller MPC architecture in order to maintain tractability. An approach is needed that decomposes the energy management problem for decentralized parallel computation, so that complex models of the overall network as well as all of the deployed devices can be handled without the problem of prohibitively long computation times. Recently, constraint tightening and Lagrangian dual decomposition have been proposed to improve the situation where non-convex MILP problems cannot be decomposed and, therefore, need to be optimized centrally, considering both finite time and performance guarantees. In particular, the use of dual decomposition and constraint tightening has been demonstrated for decentralized optimization of up to 10,000 modeled electric vehicles, each subject to shared resource constraints [55, 56]. The proposed algorithms allow parallel computation of subproblems and require only one iterative stage to achieve finite-time convergence to a feasible solution [56]. Whereas a commercial solver was unable to provide an adequate solution to the centralized MILP problem if the number of vehicles exceeded 700, the decomposition method provides solutions in 30 min for 10,000 vehicles. However, despite these attractive features, the method was not incorporated into an online control scheme. Moreover, the method was not demonstrated for the more complex application to flexible multi-energy networks, which typically requires greater computational effort due to additional constraints used to model energy system interactions and network behaviour.

In summary, the review presented in this section focussed on modeling approaches and energy management strategies applied to multi-energy systems. The two main areas which are not fully addressed in these studies are the lack of computationally efficient integrated energy network modeling approaches and poor scalability in optimization-based control methods for which computation time must be substantially reduced. The following sections present novel optimization modeling and a decentralized control approach within a multi-agent framework approach to address each of these issues [11, 12, 46].

8.3 Modeling of the Multi-Energy District

A general multi-energy system modeling framework is defined in this section. The framework takes inspiration from several MILP-based multi-energy system modeling approaches [9, 10, 21]. Specifically, the framework adopts and extends the representation of the hybrid MLD system described in [10], which is perhaps the easiest to understand in the context of multi-energy systems and should be familiar to control practitioners, as an extension of the original MLD modeling formulation [30].

The purpose of the framework is to enable an end-user to generate an optimization problem, which can then be readily integrated into an MPC scheme. Therefore, the framework should facilitate the generation of constraints and a cost function for multi-energy systems of arbitrary configuration, and also be amenable to software implementation. As in [10], this is achieved by defining a set of individual modules which represent the component parts of a multi-energy system, which can then be connected to one another in any desired configuration. Individual modules may represent conversion devices, system loads, storage or connecting networks. Each pre-defined module contains a set of modeling constraints to describe the basic physical characteristics of that component. The framework therefore provides a scalable and versatile means of MPC model development for multi-energy systems.

8.3.1 Generic Module Representation

The general representation of MILP module constraints is given as [11]

$$\boldsymbol{x}_{k+1} = \boldsymbol{A}\boldsymbol{x}_k + \boldsymbol{B}_u\boldsymbol{u}_k + \boldsymbol{B}_d\boldsymbol{\delta}_k + \boldsymbol{B}_z\boldsymbol{z}_k + \boldsymbol{B}_w\boldsymbol{w}_k \tag{8.1}$$

$$\boldsymbol{y}_k = \boldsymbol{C}\boldsymbol{x}_k + \boldsymbol{D}_u\boldsymbol{u}_k + \boldsymbol{D}_d\boldsymbol{\delta}_k + \boldsymbol{D}_z\boldsymbol{z}_k + \boldsymbol{D}_w\boldsymbol{w}_k \tag{8.2}$$

$$\boldsymbol{E}_d\boldsymbol{\delta}_k + \boldsymbol{E}_z\boldsymbol{z}_k \leq \boldsymbol{E}_u\boldsymbol{u}_k + \boldsymbol{E}_x\boldsymbol{x}_k + \boldsymbol{E}_w\boldsymbol{w}_k + \boldsymbol{E}, \tag{8.3}$$

where k is the discrete-time index, $\boldsymbol{x}_k \in \mathbb{R}^{N_x}$ is the vector of system states, $\boldsymbol{u}_k \in \mathbb{R}^{N_{uc}}$ is the vector of control inputs, $\boldsymbol{\delta}_k \in \mathbb{Z}^{N_\delta}$ is the vector of integer variables, $\boldsymbol{z}_k \in \mathbb{R}^{N_z}$ is the vector of auxiliary variables, $\boldsymbol{w}_k \in \mathbb{R}^{N_w}$ is the vector of disturbance forecasts and $\boldsymbol{y}_k \in \mathbb{R}^{N_y}$ is the vector of system outputs. The vector $\boldsymbol{E}$ contains upper and lower bounds imposed on system variables as well as fixed system inputs. Variable values are held constant during each sampling period, i.e. they are modeled with zero-order hold.

In order to accurately represent non-linear component behaviour, modules may also be defined that feature quadratic inequality constraints. Hence, the general representation of (8.1)–(8.3) is extended with the addition of constraints of the following form:

$$\frac{1}{2}\boldsymbol{z}_k^\top \boldsymbol{F}_{zz_n}\boldsymbol{z}_k + \boldsymbol{F}_{z_n}^\top \boldsymbol{z}_k \leq F_n, \quad n = 1, \ldots, m \tag{8.4}$$

in which m is the total number of quadratic constraints, $\boldsymbol{F}_{zz_n}$ is a $N_z \times N_z$ matrix, $\boldsymbol{F}_{z_n}$ is a $N_z \times 1$ vector and F_n is a scalar.

All of the components of a generic multi-energy system are represented as distinct modules which may be connected up to form larger aggregate modules. Hence, energy demand, conversion and storage components can be aggregated to represent energy systems in buildings and these in turn can be aggregated to represent building districts (see Fig. 8.2). When components are aggregated, their variable vectors are concatenated, their respective coefficient matrices in (8.1)–(8.3) are combined using block-diagonal concatenation and matrices in (8.4) are padded with zeros to ensure the correct dimensions. For example, if two components are aggregated, their continuous auxiliary variable vectors, $\boldsymbol{z}_{k_1} \in \mathbb{R}^{N_{z1}}$ and $\boldsymbol{z}_{k_2} \in \mathbb{R}^{N_{z2}}$, would be concatenated to $\boldsymbol{z}_k \in \mathbb{R}^{N_{z1}+N_{z2}}$, matrices $\boldsymbol{B}_{z1}$ and $\boldsymbol{B}_{z2}$ would be combined to form $\boldsymbol{B}_z = \begin{bmatrix} \boldsymbol{B}_{z1} & 0 \\ 0 & \boldsymbol{B}_{z2} \end{bmatrix}$ and new matrix $\boldsymbol{F}_{zz_i}^{N_z \times N_z}$ and vector $\boldsymbol{F}_{z_i}^{N_z \times 1}$ would be generated, for which $N_z = N_{z1} + N_{z2}$.

To form connections within aggregations, output variables of each component module, y_k, are designated as either source or sink *ports* for connection to the ports of other components. Connection constraints are added which define the flow of energy between connected ports:

$$y_{k,\text{snk}_j} = \gamma_{ij} \cdot y_{k,\text{src}_i} \tag{8.5}$$

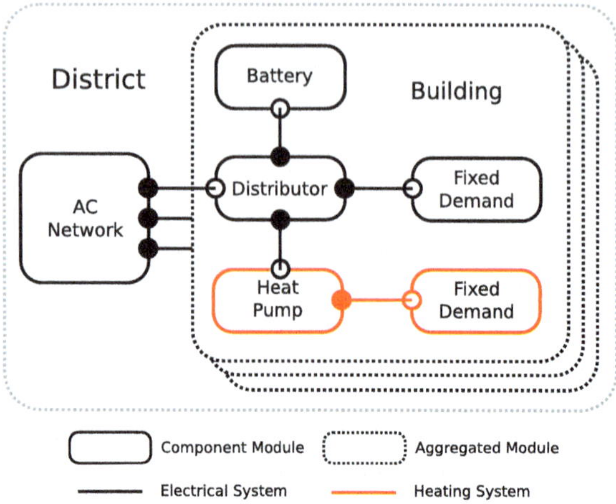

Fig. 8.2 Illustration showing the recursive aggregation of connected component modules. Multiple components are aggregated to form a building module and in turn multiple buildings are aggregated to form a district. Filled circles indicate source ports, empty circles indicate sink ports and lines indicate connections [11]

For example, the heat output of a heat pump would be designated as a source port and may be connected to the sink port of a heat demand. By convention, a positively valued output variable at a sink port, y_{k,snk_j}, always represents a physical flow into the component at that port. Furthermore, γ_{ij} may take values of any magnitude, to represent physical transmission losses ($|\gamma_{ij}| < 1$), to convert between different units of a physical quantity, e.g. between kW and p.u. of electricity, or to convert between different physical quantities of relevance, e.g. conversion between a mass flowrate of water at a given temperature and the amount of heat energy available from the water.

Whenever there is a need to connect several output ports of the same physical type at a common junction, then a *distributor* module is used with a defined number of source and sink ports:

$$y_{\text{snk}_i} = z_{\text{snk}_i} \tag{8.6}$$

$$y_{\text{src}_j} = z_{\text{src}_j} \tag{8.7}$$

$$\sum_{i=1}^{N_{\text{snk}}} z_{\text{snk}_i} - \sum_{j=1}^{N_{\text{src}}} z_{\text{src}_j} = 0, \tag{8.8}$$

where N_{snk} and N_{src} are the number of sink and source ports of the distributor, respectively. The distributor module may be aggregated and connected via connection constraints like any other module.

The benefits of this aggregation become clear when using an object-oriented software implementation of the framework. It allows optimization models to be developed in a logical, bottom-up manner, in which named component objects are first instantiated individually, before being passed as sub-components for the instantiation of aggregation objects. The properties of these aggregation objects comprise the combined properties of the sub-components, alongside a specific 'sub-components' property which stores individual instances of each sub-component object. Since aggregation objects can be created recursively, this makes interrogation of the model to find the properties of specific components easier, as successive aggregations create a hierarchy to facilitate navigation through the model. This is particularly useful for very large optimization models. It also allows an entire district to be represented by a single, top-level object, one which stores the variable names (sorted by type), constraints and cost function terms of all aggregated sub-components in a single location. This means that the constraint and cost function properties from this single object can be passed to an optimization solver at run time, reducing the programming effort required to locate and parse this information. Furthermore, when viewing the optimization results, it is possible to retrieve all variable values of a given type, e.g. all control inputs, by simply referring to the top-level object. Further details can be found in [11, 46].

8.3.2 *Energy Conversion Device Modules*

8.3.2.1 Reversible Water-Source Heat Pump Module

The prosumer-enabled bi-directional ULTDHC can deliver both heating and cooling to/from the network and therefore individual buildings are able to operate their own reversible WSHPs. The reversible WSHP model is given as follows [11]:

$$y_{k,\text{hp,elec}} = \begin{cases} 1/\text{COP} \cdot y_{k,\text{hp,out}}, & \text{(heating mode)} \\ 1/\text{EER}(-y_{k,\text{hp,in}}), & \text{(cooling mode)} \end{cases} \tag{8.9}$$

$$y_{k,\text{hp,elec}} = \left| y_{k,\text{hp,out}} - y_{k,\text{hp,in}} \right| \tag{8.10}$$

$$\text{COP} = \eta_{\text{carnot,h}} \frac{\theta_{load}^{heat}}{\theta_{load}^{heat} - (\theta_w + \theta_c)/2} \tag{8.11}$$

$$\text{EER} = \eta_{\text{carnot,c}} \frac{\theta_{out}^{cool}}{(\theta_w + \theta_c)/2 - \theta_{load}^{cool}}, \tag{8.12}$$

which account for the differences between the heating and cooling modes of the heat pump. When in heating mode, the positive heat output $y_{k,\text{hp,out}}$ is the product of the coefficient of performance (COP) and the electrical power input $y_{k,\text{hp,elec}}$. In cooling mode, the direction of heat flow is reversed, so that the product of the energy efficiency ratio (EER) and the positive electrical power input must be equal to the negative value of heat flow at the input side, $y_{k,\text{hp,in}}$. Constant COP and EER values are assumed since the optimal temperature set points in the warm and cool pipes of the ULTDHC network, θ_w and θ_c, are fixed. The required temperatures to serve heating and cooling loads are θ_{load}^{heat} and θ_{load}^{cool}, respectively. The Carnot efficiencies $\eta_{\text{carnot,h}}$ and $\eta_{\text{carnot,c}}$ represent reductions from the theoretical maximum, or Carnot, COP and EER.

Following the example of [9], these equations can be expressed as the following constraint set:

$$y_{k,\text{rwhp,elec}} = u_{k,\text{rwhp}} \tag{8.13}$$
$$y_{k,\text{rwhp,in}} = z_{k,\text{rwhp,in}} \tag{8.14}$$
$$y_{k,\text{rwhp,out}} = z_{k,\text{rwhp,out}} \tag{8.15}$$
$$z_{k,\text{rwhp,in}} = ((\text{COP} - 1) + \text{EER})z_{k,\text{rwhp,fwd}} - \text{EER}u_{k,\text{rwhp}} \tag{8.16}$$
$$z_{k,\text{rwhp,out}} = (\text{COP} + (\text{EER} + 1))z_{k,\text{rwhp,fwd}} - (\text{EER} + 1)u_{k,\text{rwhp}} \tag{8.17}$$
$$0 \leq u_{k,\text{rwhp}} \leq \overline{u}_{k,\text{rwhp}} \tag{8.18}$$
$$-\overline{z}_{k,\text{rwhp,in}} \leq z_{k,\text{rwhp,in}} \leq \overline{z}_{k,\text{rwhp,in}} \tag{8.19}$$
$$-\overline{z}_{k,\text{rwhp,out}} \leq z_{k,\text{rwhp,out}} \leq \overline{z}_{k,\text{rwhp,out}}, \tag{8.20}$$

together with the introduction of binary variable $\delta_{k,\text{rwhp,fwd}}$ and mixed-integer linear inequalities equivalent to the following logical conditions:

$$z_{k,\text{rwhp,in}} \geq 0 \iff \delta_{k,\text{rwhp,fwd}} = 1$$
$$z_{k,\text{rwhp,fwd}} = \delta_{k,\text{rwhp,fwd}} u_{k,\text{rwhp}}.$$

The outputs $y_{k,\text{rwhp,elec}}$, $y_{k,\text{rwhp,in}}$ are designated as sink ports whilst $y_{k,\text{rwhp,out}}$ is a source port.

8.3.2.2 Bi-directional Variable Speed Pump Module

Circulation pumps are required to provide a pressure gain at each ULTDHC substation. Since the pumps are electrically driven, they are also a point of coupling between the electricity and ULTDHC systems. The pressure gain $y_{k,\text{vsp}\Delta H}$ and mass flow into the pump $y_{k,\text{vsp,in}}$ are already determined by the ULTDHC network model, an equation is therefore needed to determine the electrical power, $y_{k,\text{vsp,elec}}$, used at this operating point:

$$y_{k,\text{vsp,elec}} = \frac{\rho \cdot g \left| y_{k,\text{vsp,in}} \right| y_{k,\text{vsp}\Delta H}}{\eta_m \cdot \eta_h}, \tag{8.21}$$

where the motor efficiency η_m and hydraulic efficiency η_h are assumed constant [57]. This equation is bilinear and involves a discontinuity because electrical power is consumed whether the pump is acting on a positive- or negative-valued mass flow. The discontinuity can be represented by following the example of [9] and introducing the auxiliary variables $\delta_{k,\text{vsp,fwd}}$ and $z_{k,\text{vsp,fwd}}$:

$$y_{k,\text{vsp,elec}} = z_{k,\text{vsp,elec}} \tag{8.22}$$
$$y_{k,\text{vsp,in}} = z_{k,\text{vsp,in}} \tag{8.23}$$

$$y_{k,\text{vsp,out}} = z_{k,\text{vsp,out}} \tag{8.24}$$

$$y_{k,\text{vsp},\Delta H} = z_{k,\text{vsp},\Delta H} \tag{8.25}$$

$$z_{k,\text{vsp,in}} = z_{k,\text{vsp,out}} \tag{8.26}$$

$$z_{k,\text{vsp,elec}} = 2C_{\text{vsp}} z_{k,\text{vsp,fwd}} z_{k,\text{vsp},\Delta H} - C_{\text{vsp}} z_{k,\text{vsp,in}} z_{k,\text{vsp},\Delta H} \tag{8.27}$$

where $C_{\text{vsp}} = \dfrac{\rho \cdot g}{\eta_m \cdot \eta_h}$,

together with mixed-integer linear inequalities equivalent to the following logical conditions:

$$z_{k,\text{vsp,in}} \geq 0 \iff \delta_{k,\text{vsp,fwd}} = 1 \tag{8.28}$$

$$z_{k,\text{vsp,fwd}} = \delta_{k,\text{vsp,fwd}} z_{k,\text{vsp,in}}. \tag{8.29}$$

This does not address the bilinear term in (8.21). However, an approximate McCormick relaxation can be used to replace bilinear terms and make the model convex. Moreover, when using the multi-agent approach described in Sect. 8.4, an iterative procedure is used to avoid the bilinear term altogether, by fixing the value of pump pressure head.

The outputs $y_{k,\text{vsp,elec}}$, $y_{k,\text{vsp,in}}$ are designated as sink ports whilst $y_{k,\text{vsp,out}}$ is a source port.

8.3.3 Storage Device Modules

8.3.3.1 Battery Storage Module

Batteries provide essential storage capability for electrical system flexibility but exhibit losses upon charging and discharging due to internal resistance. Losses must be modeled for both the forward and reverse flows through the battery, requiring separate charging and discharging efficiencies. The state-space battery storage model from [9] describes separate charging and discharging efficiencies, which apply to mutually exclusive charging and discharging states, represented by the following constraints:

$$x_{k+1,\text{batt}} = (1 - L_{\text{batt}}) x_{k,\text{batt}} + \left(\eta_{\text{chg}} - \frac{1}{\eta_{\text{dchg}}}\right) z_{k,\text{batt,chg}} \Delta t + \frac{1}{\eta_{\text{dchg}}} u_{k,\text{batt,chg}} \Delta t \tag{8.30}$$

$$y_{k,\text{batt,chg}} = u_{k,\text{batt,chg}} \tag{8.31}$$

$$\underline{x}_{k,\text{batt}} \leq x_{k,\text{batt}} \leq \overline{x}_{k,\text{batt}} \tag{8.32}$$

$$\underline{u}_{k,\text{batt,chg}} \leq u_{k,\text{batt,chg}} \leq \overline{u}_{k,\text{batt,chg}}, \tag{8.33}$$

with the introduction of binary variable $\delta_{k,\text{batt,chg}}$ and the following logical conditions [9, 30]:

$$u_{k,\text{batt,chg}} \geq 0 \iff \delta_{k,\text{batt,chg}} = 1 \tag{8.34}$$

$$z_{k,\text{batt,chg}} = \delta_{k,\text{batt,chg}} u_{k,\text{batt,chg}}, \tag{8.35}$$

where $x_{k,\text{batt}}$ is the amount of energy stored in the battery, $u_{k,\text{batt,chg}}$ is the control input, η_{chg} and η_{dchg} are the efficiencies of charging and discharging, respectively, and L_{batt} is the standby loss factor due to battery self-discharge. The output amount of battery charging $y_{k,\text{batt,chg}}$ is designated as a sink port.

8.3.3.2 Thermal Storage Module

Assuming a perfectly mixed water vessel and well-insulated pipes, the only losses considered from thermal storage are due to heat transfer through vessel walls to the surroundings. Therefore, a simple state-space model is used for thermal storage modules [11]:

$$x_{k+1,\text{thm}} = (1 - L_{\text{thm}})x_{k,\text{thm}} + u_{k,\text{thm,chg}}\Delta t \tag{8.36}$$

$$y_{k,\text{thm,chg}} = u_{k,\text{thm,chg}} \tag{8.37}$$

$$\underline{x}_{k,\text{thm}} \leq x_{k,\text{thm}} \leq \overline{x}_{k,\text{thm}} \tag{8.38}$$

$$\underline{u}_{k,\text{thm,chg}} \leq u_{k,\text{thm,chg}} \leq \overline{u}_{k,\text{thm,chg}}, \tag{8.39}$$

where $x_{k,\text{thm}}$ is the amount of energy stored, $u_{k,\text{thm,chg}}$ is the control input and L_{thm} is the standby loss factor for loss of energy to the surroundings. The output amount of storage charging, $y_{k,\text{thm}}$, is designated as a sink port.

8.3.4 Prosumer Modules

8.3.4.1 Fixed Generation/Demand Modules

Fixed energy generation/demand is the term given to system input/output energy flows that cannot be controlled, only forecast; they are acknowledged, or 'known', disturbances to the system [11]. These are modeled as fixed flows into or out of the system, equivalent to the current (measured) and future (forecast) values. Such an approach in an MPC setting is known as *certainty-equivalent* MPC [58], in which an uncertain optimization problem is transformed into a deterministic one by assuming that all future disturbances are equal to deterministic estimates.

The known disturbance variables $w_{k,\text{fixd},j}$ are the predicted profiles of generation/demand j from a chosen forecasting method:

$$y_{k,\text{fixd},j} = w_{k,\text{fixd},j}. \tag{8.40}$$

The output $y_{k,\text{fixd},j}$ is designated as a sink port.

8.3.5 Network Energy Flow Modules

The models of the electric and the ULTDHC networks are presented in this section [11, 12].

8.3.5.1 Electric Power Network Module

An optimal power flow problem using a second-order cone relaxation is used to accurately model an AC electric network [35]. This problem describes a balanced, radial network with a pre-specified set of bus power injections. The optimal solution determines the required power injection at the slack bus, as well as line powers, currents and bus voltages, which are constrained to be within statutory limits.

Let $\mathcal{G}^E = (\mathcal{N}^E, \mathcal{A}^E)$ be a directed graph representing a radial AC network, where nodes $\mathcal{N}^E := \{0, 1, \ldots, n\}$ represent buses in the network and each arc $\mathcal{A}^E \subseteq \mathcal{N}^E \times \mathcal{N}^E$ represents a distribution line. Denote an arc originating at node i and ending at node j as (i, j) and let R_{ij} and X_{ij} be the resistance and reactance on that line. Let $z_{k,P_{ij}}$ and $z_{k,Q_{ij}}$ be the sending end active and reactive power flows along line (i, j) and let y_{k,p_j} and y_{k,q_j} be the active and reactive power injections at each bus j. Let $z_{k,I_{ij}}$ be the complex current along line (i, j) and $z_{k,\ell_{ij}}$ the magnitude of current squared. Finally, let z_{k,V_j} be the complex voltage at each bus j and z_{k,ν_j} the magnitude of voltage squared. The root of the graph $\mathcal{G}^E$ (node 0) represents the substation bus which is connected to a higher voltage network and is given a reference voltage z_{k,ν_0} of 1 p.u. The orientation of the directed graph is chosen with all arcs directed away from the root node, as shown in Fig. 8.3.

The branch flow model for radial networks is relaxed to form the following convex model (8.41)–(8.43), $\forall j \in \mathcal{N}^E$ and (8.44), $\forall (i, j) \in \mathcal{A}^E$ [35]:

$$y_{k,p_j} = \sum_{l:j\to l} z_{k,P_{jl}} - \sum_{i:i\to j} (z_{k,P_{ij}} - R_{ij} z_{k,\ell_{ij}}) \tag{8.41}$$

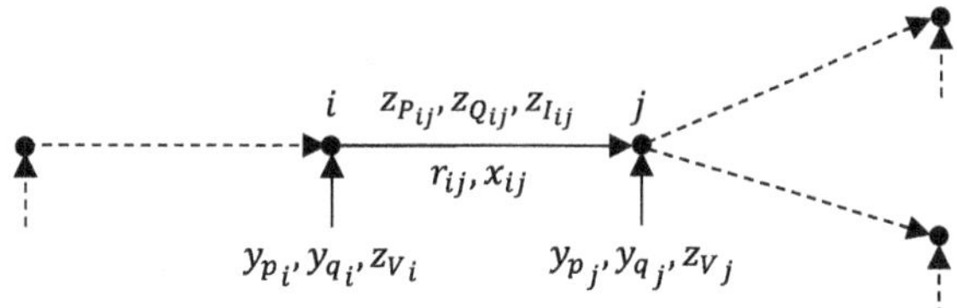

Fig. 8.3 Generic representation of AC network section [11]

$$y_{k,q_j} = \sum_{l:j\to l} z_{k,Q_{jl}} - \sum_{i:i\to j} (z_{k,Q_{ij}} - X_{ij} z_{k,\ell_{ij}}) \tag{8.42}$$

$$z_{k,v_i} - z_{k,v_j} - 2(R_{ij} z_{k,P_{ij}} + X_{ij} z_{k,Q_{ij}}) + (R_{ij}^2 + X_{ij}^2) z_{k,\ell_{ij}} = 0 \tag{8.43}$$

$$z_{k,P_{ij}}^2 + z_{k,Q_{ij}}^2 - z_{k,\ell_{ij}} z_{k,v_i} \leq 0. \tag{8.44}$$

The outputs y_{k,p_j} and y_{k,q_j} are designated as source ports.

Due to the inequality in (8.44), it is necessary to solve the problem with an objective that is strictly increasing in $z_{k,\ell_{ij}}$ in order to yield a feasible solution to the original network model [35]. A penalty cost term associated with $z_{k,\ell_{ij}}$ is therefore added to the objective function:

$$c_{k,\mathrm{AC}} = \sum_{i\to j\in A^E} \lambda_{\ell_{ij}}^{\mathrm{val}} z_{k,\ell_{ij}}, \tag{8.45}$$

with $\lambda_{\ell_{ij}}^{\mathrm{val}} > 0$, which is intended to help recover an *exact*, feasible solution from the model, i.e. when (8.44) is satisfied to equality.

8.3.5.2 ULTDHC Network Module

Several classifications of ULTDHC networks can be made [45]. These include whether the network is an open or closed system, the number of pipelines used at differing temperature levels, if the network is radial or in a ring topology and whether energy and mass flows within the network are uni- or bi-directional. Here a closed system of two pipelines is considered, one warm and one cool, with bi-directional flows between them [46]. The proposed model can be applied to networks with a radial or ring topology. Let $\mathcal{G}^W = (\mathcal{N}^W, \mathcal{A}^W)$ be a directed graph representing the warm side of the ULTDHC network, where nodes $\mathcal{N}^W := \{0, 1, \ldots, n\}$ represent junctions in the network and each arc $\mathcal{A}^W \subseteq \mathcal{N}^W \times \mathcal{N}^W$ represents a pipe. Denote an arc originating at node i and ending at node j as (i, j). The mass flow of water injected to the warm side at each junction j is y_{k,m_j}, whilst $z_{k,\mathrm{w}H_j}$ is the pressure head of water at that point in $\mathrm{m_{H_2O}}$. Pressure head at the cold side of junction j is denoted $z_{k,\mathrm{c}H_j}$. Let $z_{k,\mathrm{sub}\Delta H_j} \geq 0$ be the head loss that occurs with flow y_{k,m_j} through pressure-reducing elements of a substation located at j. The head gain supplied by the substation pump is denoted $z_{k,\mathrm{pump}\Delta H_j} \geq 0$. Let $z_{k,\mathrm{w}\Delta H_{ij}} \geq 0$ be the frictional head loss between the ends of pipe (i, j) for a given mass flow $z_{k,m_{ij}}$. By convention, head loss in a pipe is given a positive value, whilst flow $z_{k,m_{ij}}$ is considered positive in the direction from i to j (see Fig. 8.4). Another convention is defined that heating supplied to a substation from a warm pipe is given by a negative flow of warm water y_{k,m_j} at a node j, whereas rejection of heat at a substation is given by a positive nodal flow. Assuming incompressible flow and constant water properties, the equation for conservation of mass at each node $j \in \mathcal{N}^W$ (8.47) and the Darcy–Weisbach pressure drop equation in all warm pipes $(i, j) \in \mathcal{A}^W$ (8.48) can be defined [37]:

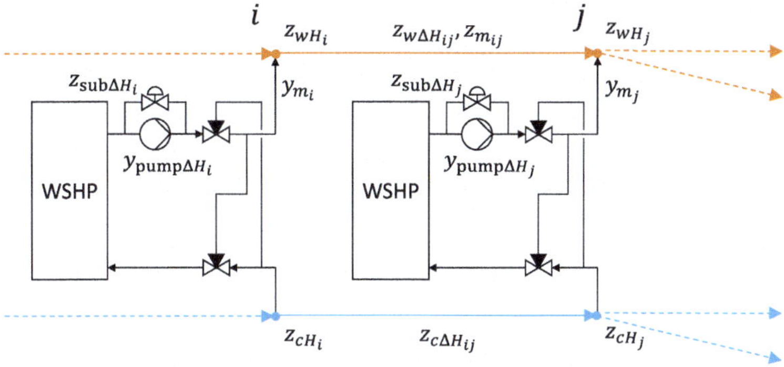

Fig. 8.4 Generic representation of ULTDHC network section [11]

$$y_{k,m_j} = z_{k,m_j} \tag{8.46}$$

$$z_{k,m_j} - \sum_{l:j\to l} z_{k,m_{jl}} + \sum_{l:l\to j} z_{k,m_{lj}} = 0 \tag{8.47}$$

$$C_{ij} z_{k,m_{ij}} \left| z_{k,m_{ij}} \right| - z_{k,\mathrm{w}H_i} + z_{k,\mathrm{w}H_j} = 0 \tag{8.48}$$

$$\text{where } C_{ij} = \frac{8L_{\text{pipe}} f}{g D^5 \rho^2 \pi^2},$$

where L_{pipe} is the pipe length in m, f is the pipe friction factor (assumed constant), $g = 9.81\ \mathrm{m}/s^2$ is gravitational acceleration, D is the internal diameter of the pipe in m and $\rho = 998\ \mathrm{kg/m^3}$ is the density of water. Constraint (8.48) is piecewise quadratic due to the bilinear term, which accounts for the direction of pressure drop. This can be reformulated in terms of absolute pressure drop:

$$C_{ij} z_{k,m_{ij}}^2 - z_{k,\mathrm{w}\Delta H_{ij}} = 0, \tag{8.49}$$

with the addition of mixed-integer linear inequalities and the binary indicator variable $\delta_{k,m_{ij}}$ to express the following logic [11, 30]:

$$z_{k,m_{ij}} \geq 0 \iff \delta_{k,m_{ij}} = 1 \tag{8.50}$$

$$z_{k,m_{ij}} \geq 0 \implies z_{k,\mathrm{w}H_i} + z_{k,\mathrm{w}H_j} = z_{k,\mathrm{w}\Delta H_{ij}} \tag{8.51}$$

$$z_{k,m_{ij}} \leq 0 \implies z_{k,\mathrm{w}H_i} + z_{k,\mathrm{w}H_j} = -z_{k,\mathrm{w}\Delta H_{ij}}, \tag{8.52}$$

where $z_{k,\mathrm{w}\Delta H_{ij}} \geq 0$ is ensured due to the squared term in (8.49).

Constraint (8.49) is no longer piecewise but it is still a non-convex quadratic equality; it must be relaxed to a convex inequality in order to be solved efficiently:

$$C_{ij} z_{k,m_{ij}}^2 - z_{k,\mathrm{w}\Delta H_{ij}} \leq 0. \tag{8.53}$$

This particular reformulation in (8.53) is chosen over comparable approaches [36, 39] because it only requires a single binary indicator variable and a single quadratic constraint to be included per pipe.

The warm and cold side pipes are coupled by mass flows and exhibit symmetry, meaning that known mass flows in the warm side are also known in the cold side [37]. If the warm and cold side pipe networks are assumed to be identical, then it follows that frictional pressure losses are also symmetrical. The following can therefore be defined $\forall(i, j) \in \mathcal{A}^W$:

$$z_{k,\mathrm{w}H_i} - z_{k,\mathrm{w}H_j} + (z_{k,\mathrm{c}H_i} - z_{k,\mathrm{c}H_j}) = 0, \tag{8.54}$$

reducing the number of constraints needed to determine pressures in the cold side.

Similarly to the pipe network, the Darcy–Weisbach equation can also be reformulated for pressure drop through individual substations, $z_{k,\mathrm{sub}\Delta H_j}$, $\forall j \in \mathcal{N}^W$:

$$y_{k,\mathrm{pump}\Delta H_j} = z_{k,\mathrm{pump}\Delta H_j} \tag{8.55}$$

$$C_j z_{k,m_j}^2 - z_{k,\mathrm{sub}\Delta H_j} \leq 0, \tag{8.56}$$

with the addition of mixed-integer linear inequalities and the binary indicator variable δ_{k,m_j} used to express the following logical conditions [11, 30]:

$$z_{k,m_j} \geq 0 \iff \delta_{k,m_j} = 1 \tag{8.57}$$

$$z_{k,m_j} \geq 0 \implies z_{k,\mathrm{c}H_j} - z_{k,\mathrm{w}H_j} = z_{k,\mathrm{sub}\Delta H_j} - z_{k,\mathrm{pump}\Delta H_j} \tag{8.58}$$

$$z_{k,m_j} \leq 0 \implies z_{k,\mathrm{c}H_j} - z_{k,\mathrm{w}H_j} = -z_{k,\mathrm{sub}\Delta H_j} + z_{k,\mathrm{pump}\Delta H_j}. \tag{8.59}$$

As previously mentioned, a direction of positive mass flow, z_{k,m_j}, is by convention designated to be from the cool side to the warm side through the substation. The pump head gain, $z_{k,\mathrm{pump}\Delta H_j}$, is always positive and the pressure drop due to pressure-reducing substation elements, $z_{k,\mathrm{sub}\Delta H_j}$, is bounded from below by the quadratic constraint (8.56). The constant C_j is determined for frictional losses which are due to static equipment in the substation, e.g. heat exchangers, fittings, etc. If $z_{k,\mathrm{sub}\Delta H_j}$ exceeds these losses to match the difference between warm and cool side pressure heads, this additional pressure drop is physically interpreted as being due to valve throttling, i.e. variably increasing frictional losses by reducing the pump bypass valve opening. If water is flowing through the substation to discharge into a pipe at greater pressure head than the inlet pipe, then $z_{k,\mathrm{pump}\Delta H_j}$ must take a non-zero value to ensure (8.56) is satisfied. Otherwise, the inequalities describing logical conditions would require $z_{k,\mathrm{sub}\Delta H_j}$ to be negative, violating (8.56).

Outputs y_{k,m_j} and $y_{k,\mathrm{pump}\Delta H_j}$ are designated as source ports for connection to other component modules.

The inclusion of quadratic inequality relaxations (8.53) and (8.56) necessitates minimization of $z_{k,\mathrm{w}\Delta H_{ij}}$ and $z_{k,\mathrm{sub}\Delta H_j}$ to ensure that feasible solutions are obtained. Another objective term is to minimize pumping power throughout the network. Due to the inclusion of pump pressure gain, $z_{k_{\mathrm{pump}\Delta H_j}}$, in logical conditions describing

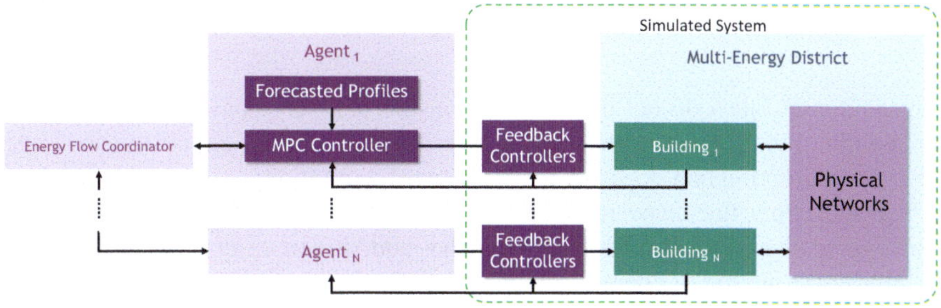

Fig. 8.5 Illustration of the multi-agent district control scheme. Although only shown for Agent 1, all agents in the scheme receive forecast data and utilize a controller [11]

substation pressure drop, minimizing $z_{k_{\mathrm{pump}\Delta H_j}}$ ensures that substation pressure drop, $z_{k,\mathrm{sub}\Delta H_j}$, is also minimized. Hence, cost terms associated with $z_{k,\mathrm{w}\Delta H_{ij}}$ and $z_{k_{\mathrm{pump}\Delta H_j}}$ are added to the ULTDHC module:

$$c_{k,\mathrm{HC}} = \sum_{i \to j \in A^W} \lambda^{\mathrm{val}}_{\Delta H_{ij}} z_{k,\mathrm{w}\Delta H_{ij}} + \sum_{j \in N^W} \lambda^{\mathrm{val}}_{\mathrm{pump}\Delta H_j} z_{k,\mathrm{pump}\Delta H_j}. \tag{8.60}$$

A feasible solution is only obtained when (8.53) is satisfied to equality, at which point the model is said to be *exact*. This takes priority over pump energy minimization and so the penalty prices are determined accordingly, with $\lambda^{\mathrm{val}}_{\Delta H_{ij}} > \lambda^{\mathrm{val}}_{\mathrm{pump}\Delta H_j} > 0$.

8.4 Decentralized Control of Multi-Energy Districts

8.4.1 Introduction

This section describes a control scheme for district energy management using **M**ulti-**A**gent dual decomposition and adaptive constraint tightening with **N**esterov's **A**ccelerated **G**radient method (MA-NAG) (Fig. 8.5) [11, 12]. A central coordinator is included, which serves a dual purpose: (1) to check global constraint violations and update dual prices; (2) to perform optimal power flow checks and ensure that the agents' requests are feasible for the operation of the multi-energy network. If there are any violations of network limits incurred by the combined agents' operating schedules then the coupling constraints are adjusted, leading to corresponding changes to the dual variables, and thus to the agents' schedules. The central coordinator is referred to as an *energy flow coordinator* in this section [11, 12].

8.4.2 Agent Control Problem

The agent controllers do not individually consider network constraints. Therefore, the coordination of agent controllers requires the use of price signals to encourage energy consumption that does not violate network limitations.

To achieve this, first an overall district problem is defined which comprises the set of all aggregated building modules, M, and has added shared capacity constraints to limit the consumption of energy from each network:

$$\begin{aligned}
\min_{v_i} \quad & \sum_{h=0}^{H-1} \sum_{i \in M} c_{k+h,i} \\
\text{s.t.} \quad & \sum_{i \in M} \boldsymbol{y}^{B}_{m_i} \leq \overline{\boldsymbol{b}}_{\mathrm{HC}}, \quad \sum_{i \in M} \boldsymbol{y}^{B}_{p_i} \leq \overline{\boldsymbol{b}}_{\mathrm{AC}}, \\
\text{and} \quad & \boldsymbol{x}_{k+1,i} = \boldsymbol{A}_i \boldsymbol{x}_{k,i} + \boldsymbol{B}_{u,i} \boldsymbol{u}_{k,i} + \boldsymbol{B}_{d,i} \boldsymbol{\delta}_{k,i} + \boldsymbol{B}_{z,i} \boldsymbol{z}_{k,i} + \boldsymbol{B}_{w,i} \boldsymbol{w}_{k,i} \qquad (\mathcal{MAP}), \\
& \boldsymbol{y}_{k,i} = \boldsymbol{C} \boldsymbol{x}_{k,i} + \boldsymbol{D}_u \boldsymbol{u}_{k,i} + \boldsymbol{D}_d \boldsymbol{\delta}_{k,i} + \boldsymbol{D}_z \boldsymbol{z}_{k,i} + \boldsymbol{D}_w \boldsymbol{w}_{k,i} \\
& \boldsymbol{E}_d \boldsymbol{\delta}_{k,i} + \boldsymbol{E}_z \boldsymbol{z}_{k,i} \leq \boldsymbol{E}_u \boldsymbol{u}_{k,i} + \boldsymbol{E}_x \boldsymbol{x}_{k,i} + \boldsymbol{E}_w \boldsymbol{w}_{k,i} + \boldsymbol{E} \\
& y_{k,\mathrm{snk}_j} = \gamma_{ij} \cdot y_{k,\mathrm{src}_i} \\
& \forall\, k \in H,\ i \in M,
\end{aligned}$$

where H is the prediction horizon length, $\overline{\boldsymbol{b}}_{\mathrm{HC}}$ is a $H \times 1$ vector of mass flow capacity limit for the ULTDHC network and $\overline{\boldsymbol{b}}_{\mathrm{AC}}$ is a $H \times 1$ vector of power capacity limit for the electricity network. The generic $\boldsymbol{y}_{m_i{}^B}$ and $\boldsymbol{y}_{p_i^B}$ are used to represent the building output connections to the respective network variables $\boldsymbol{y}_{m_i}$ and $\boldsymbol{y}_{p_i}$. The structure of $(\mathcal{MAP})$ is amenable to Lagrangian dual decomposition and the agent problems can therefore be solved in parallel with the use of a central energy flow coordinator.

By introducing $H \times 1$ dual multiplier vectors $\boldsymbol{\lambda}^{\mathrm{val}}_{\mathrm{HC}}$ and $\boldsymbol{\lambda}^{\mathrm{val}}_{\mathrm{AC}}$, the *dual problem* $(\mathcal{D})$ of the primal problem $(\mathcal{MAP})$ can be derived, in which the shared constraints are added to the objective function:

$$\begin{aligned}
\max_{\boldsymbol{\lambda}^{\mathrm{val}}_{\mathrm{HC}} \geq 0,\ \boldsymbol{\lambda}^{\mathrm{val}}_{\mathrm{AC}} \geq 0} \quad & -{\boldsymbol{\lambda}^{\mathrm{val}}_{\mathrm{HC}}}^{\top} \overline{\boldsymbol{b}}_{\mathrm{HC}} - {\boldsymbol{\lambda}^{\mathrm{val}}_{\mathrm{AC}}}^{\top} \overline{\boldsymbol{b}}_{\mathrm{AC}} \\
& + \sum_{i \in M} \min_{v_i} \left(\sum_{h=0}^{H-1} c_{k+h,i} + {\boldsymbol{\lambda}^{\mathrm{val}}_{\mathrm{HC}}}^{\top} \boldsymbol{y}^{B}_{m_i} + {\boldsymbol{\lambda}^{\mathrm{val}}_{\mathrm{AC}}}^{\top} \boldsymbol{y}^{B}_{p_i} \right). \qquad (\mathcal{D})
\end{aligned}$$

The inner minimizations within $\mathcal{D}$ are separable, lower dimensional sub-problems associated with each aggregated module i and can be solved in parallel. Algorithm 1 provides a method to exploit this property and find a good feasible solution to $(\mathcal{MAP})$ by coordinating individual agent controllers. The centralized energy flow coordinator carries out the steps in Algorithm 1, as well as the sub-routines 3 and 4, given in the appendix, whilst Algorithm 2 (see appendix) is carried out by each individual building agent. Therefore, the agents solve their local optimization problem (8.4.2)

Algorithm 1 Energy flow coordination [11]

1: Initialise $\boldsymbol{\lambda}_{\mathrm{HC}}^{\mathrm{val}}, \boldsymbol{\lambda}_{\mathrm{AC}}^{\mathrm{val}}, \boldsymbol{\mu}_{\mathrm{HC}}$ and $\boldsymbol{\mu}_{\mathrm{AC}}$ with zero vector, $\mathbf{0}^{H\times 1}$
2: Initialise $\overline{s}_{\mathrm{HC}_i}(0) = \overline{s}_{\mathrm{AC}_i}(0) = -\infty^{H\times 1}, \quad i \in M$
3: Initialise $\underline{s}_{\mathrm{HC}_i}(0) = \underline{s}_{\mathrm{AC}_i}(0) = \infty^{H\times 1}, \quad i \in M$
4: $j = 0, n = 0$
5: **if** new candidate solutions, $\boldsymbol{v}_i$, received from all agents $i \in M$ **then**
6: $\quad \overline{s}_{\mathrm{HC}_i}(j+1) = \max\{\overline{s}_{\mathrm{HC}_i}(j), \boldsymbol{y}_{m_i}^B(j+1)\},\ i \in M$
7: $\quad \underline{s}_{\mathrm{HC}_i}(j+1) = \min\{\underline{s}_{\mathrm{HC}_i}(j), \boldsymbol{y}_{m_i}^B(j+1)\},\ i \in M$
8: $\quad \boldsymbol{\rho}_{\mathrm{HC}_i}(j+1) = \overline{s}_{\mathrm{HC}_i}(j+1) - \underline{s}_{\mathrm{HC}_i}(j+1),\ i \in M$
9: $\quad \boldsymbol{\rho}_{\mathrm{HC}}(j+1) = \max(\sum_{i\in\hat{M},|\hat{M}|=H} \rho_i(j+1))$
10: Repeat 6 to 9 for AC network parameters.
11: **if** $j < 2$ OR $\sum_{i\in M} \boldsymbol{y}_{m_i}^B(j+1) \le \boldsymbol{b}_{\mathrm{HC}}$ **then**
12: Run Algorithm 3.
13: **end if**
14: **if** $j < 2$ OR $\sum_{i\in M} \boldsymbol{y}_{p_i}^B(j+1) \le \boldsymbol{b}_{\mathrm{AC}}$ **then**
15: Run Algorithm 4.
16: **end if**
17: $\nabla \boldsymbol{g}_{\mathrm{HC}}(j+1) = \sum_{i\in M} \boldsymbol{y}_{m_i}^B(j+1) - \boldsymbol{b}_{\mathrm{HC}}(j+1) + \boldsymbol{\rho}_{\mathrm{HC}}(j+1)$
18: $\nabla \boldsymbol{g}_{\mathrm{AC}}(j+1) = \sum_{i\in M} \boldsymbol{y}_{m_i}^B(j+1) - \boldsymbol{b}_{\mathrm{AC}}(j+1) + \boldsymbol{\rho}_{\mathrm{AC}}(j+1)$
19: $\boldsymbol{\mu}_{\mathrm{HC}}(j+1) = \boldsymbol{\lambda}_{\mathrm{HC}}(j) + \alpha_{\mathrm{HC}}\nabla \boldsymbol{g}_{\mathrm{HC}}(j+1)$
20: $\boldsymbol{\lambda}_{\mathrm{HC}}(j+1) = \left[\boldsymbol{\mu}_{\mathrm{HC}}(j+1) + \frac{n}{n+\beta}(\boldsymbol{\mu}_{\mathrm{HC}}(j+1) - \boldsymbol{\mu}_{\mathrm{HC}}(j))\right]_+$
21: $\boldsymbol{\mu}_{\mathrm{AC}}(j+1) = \boldsymbol{\lambda}_{\mathrm{AC}}(j) + \alpha_{\mathrm{AC}}\nabla \boldsymbol{g}_{\mathrm{AC}}(j+1)$
22: $\boldsymbol{\lambda}_{\mathrm{AC}}(j+1) = \left[\boldsymbol{\mu}_{\mathrm{AC}}(j+1) + \frac{n}{n+\beta}(\boldsymbol{\mu}_{\mathrm{AC}}(j+1) - \boldsymbol{\mu}_{\mathrm{AC}}(j))\right]_+$
23: **if** stopping criteria not met **then**
24: Broadcast $\boldsymbol{\lambda}_{\mathrm{HC}}^{\mathrm{val}}, \boldsymbol{\lambda}_{\mathrm{AC}}^{\mathrm{val}}$ and $\boldsymbol{y}_{\mathrm{cp}_i,\Delta H}$ to agents $i \in M$ with instruction "re-optimize"
25: $j = j + 1$
26: **else**
27: Shift algorithm parameter vectors $\boldsymbol{b}_{\mathrm{HC}}, \boldsymbol{b}_{\mathrm{AC}}, \boldsymbol{\lambda}_{\mathrm{HC}}^{\mathrm{val}}, \boldsymbol{\lambda}_{\mathrm{AC}}^{\mathrm{val}}, \boldsymbol{\mu}_{\mathrm{HC}}, \boldsymbol{\mu}_{\mathrm{AC}}$ and $\boldsymbol{y}_{\mathrm{cp}_i,\Delta H}$ by one time step
28: Broadcast $\boldsymbol{\lambda}_{\mathrm{HC}}^{\mathrm{val}}, \boldsymbol{\lambda}_{\mathrm{AC}}^{\mathrm{val}}$ and $\boldsymbol{y}_{\mathrm{cp}_i,\Delta H}$ to agents $i \in M$ with instruction "continue"
29: $\overline{s}_{\mathrm{HC}_i}(0) = \overline{s}_{\mathrm{AC}_i}(0) = -\infty^{H\times 1}, \quad i \in M$
30: $\underline{s}_{\mathrm{HC}_i}(0) = \underline{s}_{\mathrm{AC}_i}(0) = \infty^{H\times 1}, \quad i \in M$
31: $j = 0$
32: **end if**
33: **end if**
34: Go to step 5

and the only optimization problems solved by the coordinator are network optimal energy flow problems (8.4.2) and (8.4.2).

The inner minimizations of $\mathcal{D}$ become the local optimization problems (8.4.2) solved by each agent $i \in M$ in parallel:

$$\min_{v_i} \quad \sum_{h=0}^{H-1} c_{k+h,i} + \boldsymbol{\lambda}_{\mathrm{HC}}^{\mathrm{val}\top} \boldsymbol{y}_{m_i}^B + \boldsymbol{\lambda}_{\mathrm{AC}}^{\mathrm{val}\top} \boldsymbol{y}_{p_i}^B \tag{$\mathcal{AP}$}$$

$$\text{s.t.} \quad (1)-(3), (5) \text{ for aggregated module } i$$

The energy flow problem for the ULTDHC network includes a nodal mass flow slack variable in the cost function:

$$\begin{aligned} \min_{v} \quad & c_{k,\mathrm{HC}} + \sum_{i \in M} \lambda^{\mathrm{ref}}_{k,m_i} \cdot \left| y_{k,m_i} - r_{k,m_i} \right| \ \forall k \\ \text{s.t.} \quad & (1) - (5) \text{ for ULTDHC component.} \end{aligned} \tag{$\mathcal{HC}$}$$

The cost function $c_{k,\mathrm{HC}}$ is given in Sect. 8.3.5.2. The reference signal, r_{k,m_i}, for each nodal mass flow, y_{k,m_i}, is that determined by the agent controller located at node i, with an associated cost, $\lambda^{\mathrm{ref}}_{k,m_i}$, which is much larger than other penalty terms in $c_{k,\mathrm{HC}}$. Solving this problem determines a feasible energy flow solution with values of nodal mass flow, which are of minimum absolute distance from the reference values. If the distance is non-zero ($\left| y_{k,m_i} - r_{k,m_i} \right| > 0$) for a given sampling instance k in the prediction horizon, then the value of $b_{k,\mathrm{HC}}$ is updated to equal the sum of all mass flows determined in the energy flow solution, y_{k,m_i}; otherwise $b_{k,\mathrm{HC}}$ remains unchanged (see Algorithm 3 in the appendix). This tightening of shared constraints, by reducing $b_{k,\mathrm{HC}}$ for the affected sampling periods, increases the associated dual prices so that all agents are encouraged to shift their demand to alternative sampling intervals. The eventual result of this constraint tightening is that agents converge to local energy schedules that do not violate network limits. At this point, the stopping criteria has been met and the coordinator allows agents to implement their solutions. A secondary purpose of solving (8.4.2) is that the required pump pressure gain at each node, $\hat{y}_{k,\mathrm{cp}_i,\Delta H}$, can be communicated to individual agents and used to fix the value of $y_{k,\mathrm{cp},\Delta H}$.

Similarly, the power flow problem solved by the energy flow coordinator for the AC distribution network includes a nodal power injection slack term in the cost function:

$$\begin{aligned} \min_{v} \quad & c_{k,\mathrm{AC}} + \sum_{i \in M} \lambda^{\mathrm{ref}}_{k,p_i} \cdot \left| y_{k,p_i} - r_{k,p_i} \right| \ \forall k \\ \text{s.t.} \quad & (1)\text{–}(5) \text{ for AC component.} \end{aligned} \tag{$\mathcal{AC}$}$$

The cost function $c_{k,\mathrm{AC}}$ is given in Sect. 8.3.5.1. The total capacity limit, $\boldsymbol{b}_{\mathrm{AC}}$, for the AC network is similarly updated if $\left| y_{k,p_i} - r_{k,p_i} \right| > 0$, otherwise $\boldsymbol{b}_{\mathrm{AC}}$ remains unchanged (see Algorithm 4 in the appendix).

In both Algorithm 2 and Algorithm 1, the dual update vectors $\boldsymbol{\lambda}^{\mathrm{val}}_{\mathrm{HC}}$, $\boldsymbol{\lambda}^{\mathrm{val}}_{\mathrm{AC}}$, $\boldsymbol{\mu}_{\mathrm{HC}}$ and $\boldsymbol{\mu}_{\mathrm{AC}}$ are initialized with all zero vectors only once before the control scheme is online. Once online, these vectors are warm started at each k, determined by applying a circular shift to the final vectors of the previous control instance (step 27 of Algorithm 1). The circular shift is represented by multiplication with the matrix, $\boldsymbol{C}_{\mathrm{shift}} = \begin{bmatrix} 0 & \mathbf{I} \\ 1 & 0 \end{bmatrix}$. After applying this shift, the final element of each vector is given a zero value. This is to reduce the number of algorithm iterations taken to find a feasible solution at each sampling instance. Algorithm iterations for a given control sampling instance are interrupted once a suitable stopping criteria has been met, e.g. the agent solutions do not violate any network constraints.

In the following section of the case study, the performance of a multi-agent approach will be compared with that of an equivalent centralized MPC solution, where the latter addresses the global issue of considering all buildings and network constraints within the area simultaneously using a single controller.

8.5 Numerical Evaluation

A case study is presented in this section, in which the operation of a large multi-energy district under centralized control is simulated [11]. The district topology is based on a MATPOWER test case, 'case85' [59], and the building energy system devices reported for ULTDHC networks in [29, 60]. The district consists of an 85 bus AC distribution network and an 84 node ULTDHC network, connecting $N_B = 84$ large residential and commercial buildings/plants. The topology of the AC network is shown in Fig. 8.6, in which bus 1 represents a connection to a larger network. The topology of the ULTDHC network (also Fig. 8.6) follows that of the AC network but starts at node 2, a plant substation, where a fixed amount of waste heat is continuously supplied by a sewage treatment plant. A large borehole storage field is also located local to the sewage plant at node 2. Nodes 3 to 85 represent buildings, all of which have fixed electricity and heating demands, with the exception of nodes 31 and 59, which both have fixed cooling demands. Every building has the use of a WSHP, circulation pump, battery and thermal storage water tank (Fig. 8.7). Two circulation pumps are located at the plant substation, one for network circulation and one for pumping through the borehole storage field (Fig. 8.8). Two main branches can be identified for each network, which split at node 8. Capacity constraints and associated dual multipliers were applied separately to the sets of nodes describing these branches, which each include the nodes preceding node 8. This is to avoid constraint violations on a given branch causing constraint tightening on unaffected branches. The stopping criteria for Algorithm 1 is that, for the current iteration, feasibility checks have been carried out without adjustment to any capacity constraints and that these constraints are satisfied for both networks. Algorithm parameters used for the multi-agent scheme during simulation were $\alpha_{\text{HC}} = 1 \times 10^{-4}$, $\alpha_{\text{AC}} = 1 \times 10^{-5}$ and $\beta = 3$. When simulating the multi-agent scheme, prediction horizons of $H = 3$ and $H = 24$ are used for both a direct comparison to the centralized benchmark and to demonstrate scalability to larger problems. The penalty costs used to obtain exact solutions in the AC and ULTDHC networks are $\lambda^{\text{val}}_{\ell_{ij}} = 0.1$, $\lambda^{\text{val}}_{\Delta H_{ij}} = 1$ and $\lambda^{\text{val}}_{\text{pump}\Delta H_j} = 0.1$.

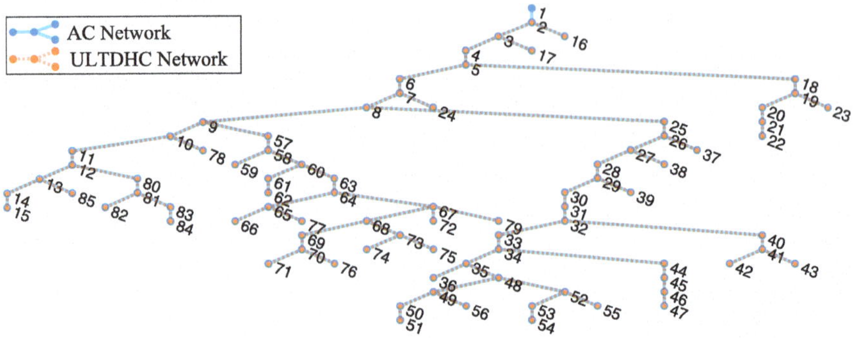

Fig. 8.6 Graphs of AC distribution network, based on MATPOWER test case 'case85' [59], and ULTDHC network [11]

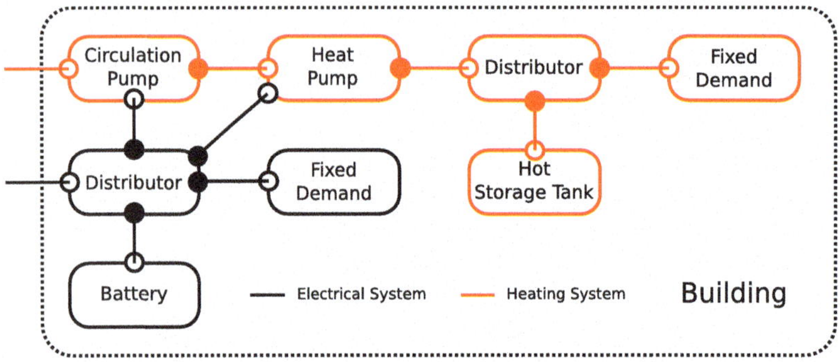

Fig. 8.7 Illustration of aggregated building module. Buildings at nodes 31 and 59 utilize a cooling system, with a cold storage tank replacing the hot storage tank shown [11]

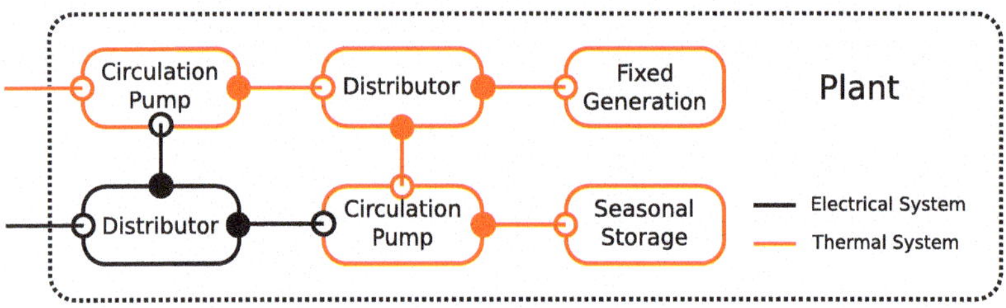

Fig. 8.8 Illustration of aggregated plant module [11]

8.5.1 *Agent Control Problem*

The objective function $J_{k,i}$ minimized by agent controller i at each time step varies between the plant substation at node 2 and the buildings at the remaining nodes. Hence, they are given separately below. For each building, the economic cost for importing, and revenue for exporting, power from/to the AC power network, as well as the cost associated with battery degradation is included in the objective function. The cost for purchasing power is the only economic term included for the plant substation objective function. A purchase and sale cost modifier is applied to the electricity distributor output variable, $y_{k,\mathrm{snk}_{1,elec}}$, of all buildings, with associated prices $\lambda^{\mathrm{buy}}_{k,y_{\mathrm{snk}_1,\mathrm{elec}}}$ and $\lambda^{\mathrm{sell}}_{k,y_{\mathrm{snk}_1,\mathrm{elec}}}$. A value cost modifier is applied to the electricity distributor output variable, $y_{\mathrm{snk}_{1,elec}}$, of the plant substation, with associated cost $\lambda^{\mathrm{val}}_{k,y_{\mathrm{snk}_1,\mathrm{elec}}}$. A battery degradation cost coefficient $\lambda^{\mathrm{abs}}_{k,u_{\mathrm{batt,chg}}}$ is applied to all batteries by adding an absolute value cost modifier applied to $u_{k,\mathrm{batt,chg}}$. The objective function for all agents includes dual variable terms associated with both the AC power and the ULTDHC networks.

The control objective for the plant substation agent is as follows:

$$J_{k,2} = \sum_{h=0}^{H-1} c_{k+h,2} + \boldsymbol{\lambda}^{\mathrm{val}\,\top}_{k+h,\mathrm{dual}} A_2 \boldsymbol{v}_{k+h,2} \tag{8.61}$$

$$c_{k+h,2} = \lambda^{\mathrm{val}}_{k,y_{2,\mathrm{snk}_1,\mathrm{elec}}} \cdot y_{k,2,\mathrm{snk}_{1,elec}} \tag{8.62}$$

$$\boldsymbol{\lambda}^{\mathrm{val}\,\top}_{k+h,\mathrm{dual}} A_2 \boldsymbol{v}_{k+h,2} = \lambda^{\mathrm{val}}_{k,y_{\mathrm{snk}_1,\mathrm{elec,dual}}} \cdot y_{k,2,\mathrm{snk}_{1,elec}} + \lambda^{\mathrm{val}}_{k,y_{\mathrm{vsp,in}}} \cdot y_{k,2,\mathrm{vsp,in}}. \tag{8.63}$$

Similarly, the control objective for the building agents is as follows:

$$J_{k,i\neq 1,2} = \sum_{h=0}^{H-1} c_{k+h,i\neq 1,2} + \boldsymbol{\lambda}^{\mathrm{val}\,\top}_{k+h,\mathrm{dual}} A_{i\neq 1,2} \boldsymbol{v}_{k+h,i\neq 1,2} \tag{8.64}$$

$$\begin{aligned} c_{k+h,i\neq 1,2} = &\left(\lambda^{\mathrm{buy}}_{k,y_{i,\mathrm{snk}_1,\mathrm{elec}}} - \lambda^{\mathrm{sell}}_{k,y_{i,\mathrm{snk}_1,\mathrm{elec}}}\right) z_{k,y_{i,\mathrm{snk}_1,\mathrm{elec}},\mathrm{buy}} + \lambda^{\mathrm{sell}}_{k,y_{i,\mathrm{snk}_1,\mathrm{elec}}} \cdot y_{k,i,\mathrm{snk}_1,\mathrm{elec}} \\ &+ \lambda^{\mathrm{abs}}_{k,u_{\mathrm{batt}_i,\mathrm{chg}}} \cdot z_{k,u_{\mathrm{batt}_i,\mathrm{chg}},\mathrm{abs}} \end{aligned} \tag{8.65}$$

$$\boldsymbol{\lambda}^{\mathrm{val}\,\top}_{k+h,\mathrm{dual}} A_{i\neq 1,2} \boldsymbol{v}_{k+h,i\neq 1,2} = \lambda^{\mathrm{val}}_{k,y_{\mathrm{snk}_1,\mathrm{elec,dual}}} \cdot y_{k,i,\mathrm{snk}_{1,elec}} + \lambda^{\mathrm{val}}_{k,y_{\mathrm{vsp,in}}} \cdot y_{k,i,\mathrm{vsp,in}}. \tag{8.66}$$

The control problem for each agent is formally stated as

$$\begin{aligned} \min_{v_i} J_{k,i} &= \sum_{h=0}^{H-1} c_{k+h,i} + \boldsymbol{\lambda}^{\mathrm{val}\,\top}_{k+h,\mathrm{dual}} A_i \boldsymbol{v}_{k+h,i} \\ \text{s.t. } &(1)\text{–}(3)\ \forall\, m_i, \end{aligned} \tag{8.67}$$

for which the m_i aggregated components are shown in Figs. 8.7 and 8.8, and the specific constraints for these components are given in Sect. 8.3.

8.5.2 Simulation Parameters and Inputs

The AC distribution network parameters are taken from a MATPOWER test case 'case85' [59] and the ULTDHC network parameters have been derived by normalizing the AC line resistance values to obtain relative distances and scaling for a pipe pressure drop of $0.03\ mH_2O/100$ m at a base mass flow rate of 10 kg/s [11]. Frictional losses in each substation are determined with the same head loss coefficient, $C_j = 5 \times 10^{-4}\ mH_2O \cdot s/kg$. Component model parameters are given in Table 8.3 and various operating limits for the district are given in Table 8.4. Data to represent a daily active power and heating demands from the 83 buildings at 15 min intervals (Figs. 8.9 and 8.10) has been assembled from an OpenEI dataset [61]. Data has been scaled so that network voltage and pressure constraints become active during simulation. For simplicity, the reactive power consumption at each node is assumed

Table 8.3 Component model coefficients

Parameter	Value[1]
Heat pump COP	3.90
Heat pump EER	2.10
Battery cycle efficiency	[0.68;0.81]
Battery standby loss factor	2.50×10^{-4}
Tank standby loss factor	2.60×10^{-4}
Borehole standby loss factor	2.50×10^{-4}
Circulation pump efficiency	0.49

[1] [*a*;*b*]—uniform random distribution

Table 8.4 District parameters

Device parameter	Value[1]	Units
AC substation capacity	3	MW
AC voltage limits	0.9 / 1.1	p.u.
ULTDHC pipe temperatures	16 / 20	°C
Plant heat input	1.60	MW
Plant pump max. head gain	50	mH_2O
Plant pump max. mass flow	250	kg/s
Borehole storage capacity	0.00 / 500	MWh
Borehole heat transfer rate	−4.00 / 4.00	MW
Borehole pump head gain	30	mH_2O
Battery capacity limits	[4.70;4.90] / [18.8;19.6]	kWh
Battery charge rate	[−12.5;−11.5] / [11.5;12.5]	kW
Tank heat capacity limits	0.00/40.0	kWh
Tank heat transfer rate	−20.0/20.0	kW
Building pump max. head gain	10	mH_2O

[1] [*a*;*b*] - uniform random distribution

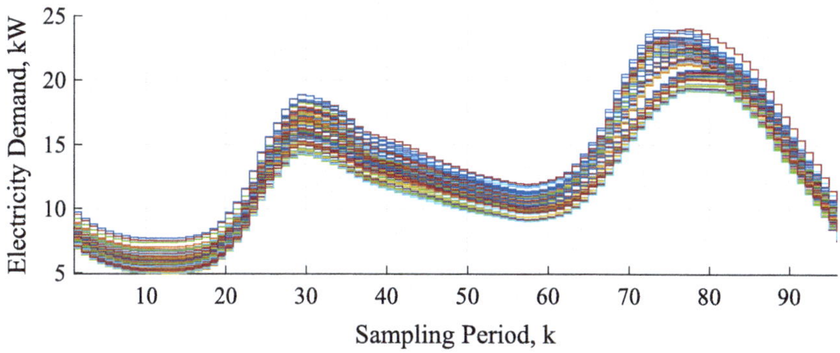

Fig. 8.9 Electricity demand profiles used in simulations [11]

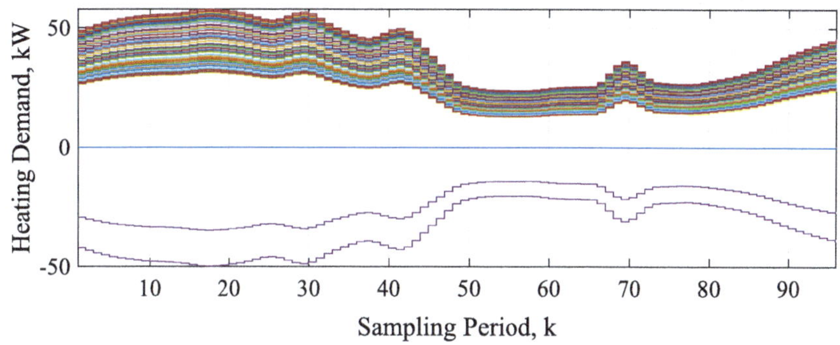

Fig. 8.10 Heating demand profiles used in simulations [11]

to be 0 p.u.; therefore, reactive power is only supplied to balance power dissipated in network lines. Electricity purchase prices have been interpolated at 15 min intervals from historic price data available from UK energy supplier Octopus Energy Ltd. for their Agile Octopus tariff [62] (see the orange plot in Figs. 8.11 and 8.12). Fixed prices are assumed for selling electricity and for battery degradation, at 1p/kWh and 3.15p/kWh, respectively. The penalty costs used to obtain exact solutions in the AC and ULTDHC networks are $\lambda_{\ell_{ij}}^{\text{val}} = 10$, $\lambda_{\Delta H_{ij}}^{\text{val}} = 100$ and $\lambda_{\text{pump}\Delta H_j}^{\text{val}} = 10$.

8.5.3 Results

The simulations were carried out using Gurobi [63] and YALMIP [64] in MATLAB [65] on a laptop with Intel Core i5-6200U CPU @ 2.30GHz and 8 GB RAM. There is little difference in the actions determined by the two schemes for a prediction horizon of $H = 3$, when comparing the total electrical and heat power supplied to the buildings (Figs. 8.11 and 8.12). This is also reflected in the very similar overall

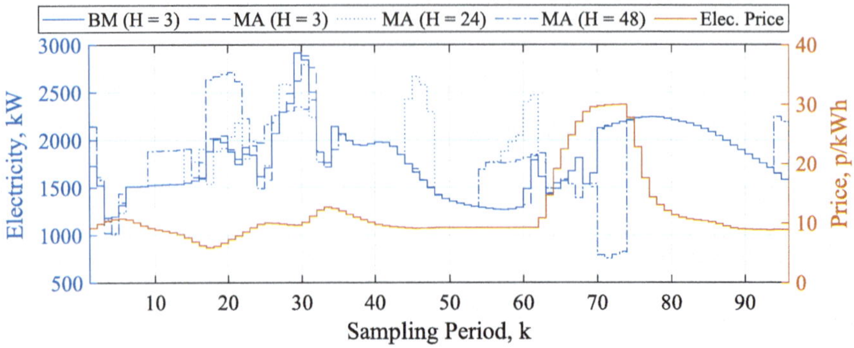

Fig. 8.11 Total electrical power supplied to buildings under benchmark (BM) and multi-agent NAG (MA) schemes [11]

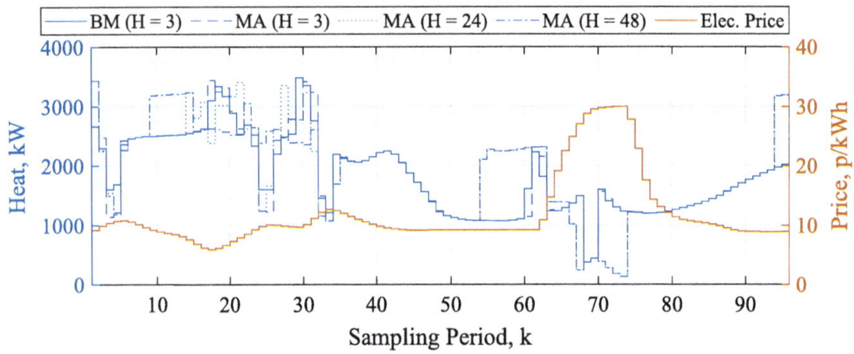

Fig. 8.12 Total heat power supplied to buildings under benchmark (BM) and multi-agent NAG (MA) schemes [11]

Table 8.5 Overall cost for purchase/sale of electricity and battery usage over 24 h period

	Benchmark	Multi-agent		
	H = 3	H = 3	H = 24	H = 48
Overall cost, £	5,247	5,265 (+0.3%)	5,121 (−2.4%)	5,102 (−2.8%)

costs to plant and building owners when using each scheme (Table 8.5). However, there is a marked difference in the computational performance (Table 8.6), with a single optimization problem in the centralized scheme taking as long as 55 min to solve and an average computation time of over 2.5 min, whereas the multi-agent equivalent scheme takes no longer than 1.01 s.

As expected, when the prediction horizon is increased to $H = 24$ and $H = 48$, the agent controllers consume energy more strategically. Greater amounts of energy are consumed during periods of lower electricity purchase price and less is consumed during price peaks. This results in a slight reduction in cost versus the centralized

Table 8.6 Computational performance over 96 sampling instances

	Benchmark	Multi-agent*		
	H = 3	H = 3	H = 24	H = 48
Mean computation time (s)	153.88	0.19	2.82	2.37
Min. computation time (s)	13.58	0.13	1.39	0.31
Max. computation time (s)	3,354.40	1.01	35.43	60.21
Std. dev. computation time (s)	464.57	0.10	3.09	7.22
Mean algorithm iterations	–	2.95	7.81	6.92
Max. algorithm iterations	–	65	121	160

*Parallel computation time for all agents is assumed to be equal to the slowest agent; communication time between agents and coordinator is assumed to be insignificant. Computation time includes that of the energy flow coordinator

benchmark. However, the main benefit of this multi-agent scheme is that, despite a 16-fold increase in the problem size, the computation time remains several orders of magnitude faster than the centralized scheme. Finally, whilst agents did often generate energy schedules which, according to detailed energy flow analysis, initially violated network constraints, it was always possible to coordinate their actions to find feasible alternatives. The case study therefore demonstrates the capability of the multi-agent control scheme to coordinate a multi-energy district without neglecting local network constraints.

In order to give greater confidence that MA-NAG can produce consistently good computational performance, a longer simulation was undertaken consisting of 2,016 sampling intervals of 15 min, i.e. a simulation length of 21 d, with $H = 24$, a prediction horizon of 4 h. This was not carried out for the benchmark scheme, due to the excessively slow computation time already highlighted, and therefore cost comparisons cannot be made. However, it is clear from the number of algorithm iterations for this extended dataset that the computational performance remains consistent (Fig. 8.13). The mean number of iterations was 6.58 and the minimum number was 2. The maximum number of iterations was 165, with more iterations occurring at the start of each day at around 4.00 am, 4 h before the morning electricity and heating load peaks, the time when both networks are most congested.

Despite these impressive results for the control of this multi-energy district, there are some limitations to the control scheme. Whilst the energy flow coordinator performs a vital role in validating the combined operating schedules of agents in the district, much like the 'independent platform' in [25], it does have the potential to

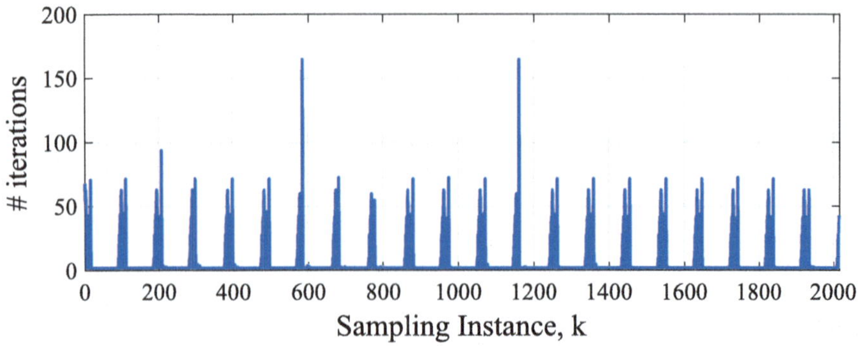

Fig. 8.13 Required number of iterations at each sampling instance for an extended 3-week simulation, $H = 24$ [11]

become a bottleneck within the scheme. This could take place either as a result of increases to the geographical scale of the district, since the sizes of optimal energy flow problems solved by the coordinator are directly proportional to the size of the networks, or due to the communication burden as every agent must communicate with the energy flow coordinator. On the other hand, there is an upper limit to the size of networks that may be considered, as pointed out in [54]. Another limitation is that the energy flow coordinator represents a single point of failure for the scheme, something which could also be said of a centralized scheme but might be avoided with a fully distributed scheme; this should be the subject of further research.

In this section, the results of the case study demonstrate that the combination of the ULTDHC modeling module and the accelerated multi-agent optimization algorithm MA-NAG is a powerful tool for coordinating large-scale district energy systems. The introduction of an energy flow coordinator enables the physical consideration of non-linear flows across multiple district energy networks. In addition, separating network security from individual energy systems addresses privacy concerns.

8.6 Conclusions and Future Work

In this chapter, it has been shown how an effective optimization-based predictive control framework for multi-energy energy communities or districts can be devised, which addresses the challenges of handling the interdependencies and interconnection of different types of energy networks, and their operation, as well as of managing heterogeneous resources across these networks, with their diverse operating modes. Realistic case studies have demonstrated the potential benefits the proposed control framework can bring to the optimal coordination of a multi-energy resources and buildings and their constrained supply networks, in terms of substantial cost savings and enhanced reliability. The designed decentralized MPC framework is able to achieve high comparable performance than a centralized approach but with a substantial decrease in computational times (e.g. reducing the maximum computation time of around 55 min for a centralized MPC controller to just over 1 s). Further work includes the control design under uncertainty and the analysis of the decentralized algorithm convergence under uncertainty, as well as the design of a fully distributed approach, without any central coordinator, which could result in a single point of failure and communication burden.

Appendix

Algorithm 2 Agent receding horizon scheme [11]

1: $t = 0,\ \Delta t > 0$
2: Initialise $\boldsymbol{\lambda}_{\text{HC}}^{\text{val}}$, $\boldsymbol{\lambda}_{\text{AC}}^{\text{val}}$ and $\boldsymbol{y}_{\text{cp}_i,\Delta H}(i \in M)$ with zero vector, $\mathbf{0}^{H\times 1}$
3: **for** $k = 0, 1, 2, \ldots$ **do**
4: Update current system states, $\mathbf{x}_k$
5: Update price forecasts, $\boldsymbol{\lambda}_{k+h}^{\text{val}}$, $h = 0, \ldots, H-1$
6: Update disturbance forecasts, $\mathbf{w}_{k+h}$, $h = 0, \ldots, H-1$
7: Solve local optimization problem ($\mathcal{AP}$)
8: Communicate candidate solution, $\boldsymbol{v}_i$, to energy flow coordinator
9: Receive instruction from energy flow coordinator
10: **if** instruction == "re-optimize" **then**
11: Go to step 7
12: **end if**
13: Pass control inputs for next sampling interval, $\mathbf{u}_k$, to lower-level feedback controllers; discard $\mathbf{u}_{k+h}$, $h = 1, \ldots, H-1$
14: **Wait until** $t = (k+1)\Delta t$
15: $k = k + 1$
16: **end for**

Algorithm 3 Sub-routine for checking ULTDHC network solution feasibility [11]

for $i = 1$ to $|M|$ **do**
    $\boldsymbol{r}_{m_i} = \boldsymbol{y}^B_{m_i}(j+1)$
end for
Solve $(\mathcal{HC})$
for $i = 1$ to $|M|$ **do**
    $\hat{\boldsymbol{y}}_{\text{cp}_i,\Delta H} = \boldsymbol{y}_{\text{pump}\Delta H_i}$
    if $\left|y_{k,m_i} - r_{k,m_i}\right| > 0 \quad \forall k$ **then**
        $b_{k,\text{HC}} = \sum_{i \in M} y_{k,m_i}(j+1)$
    end if
end for
$n = 0$

Algorithm 4 Sub-routine for checking AC network solution feasibility [11]

for $i = 1$ to $|M|$ **do**
    $\boldsymbol{r}_{p_i} = \boldsymbol{y}^B_{p_i}(j+1)$
end for
Solve $(\mathcal{AC})$
for $i = 1$ to $|M|$ **do**
    if $\left|y_{k,p_i} - r_{k,m_i}\right| > 0 \quad \forall k$ **then**
        $b_{k,\text{AC}} = \sum_{i \in M} y_{k,p_i}(j+1)$
    end if
end for
$n = 0$

References

1. B.M. Weedy, B.J. Cory, N. Jenkins, J.B. Ekanayake, G. Strbac, *Electric Power Systems*, 5th edn., vol. 14 (Wiley, 2012), p. 512. ISBN: 978-0-470-68268-5
2. N. Daina, A. Sivakumar, J.W. Polak, Electric vehicle charging choices: modelling and implications for smart charging services. Transp. Res. Part C: Emerg. Technol. **81**, 36–56 (2017). ISSN: 0968-090X. https://doi.org/10.1016/j.trc.2017.05.006. [Online]. Available: https://www.sciencedirect.com/science/article/pii/S0968090X17301365
3. S. Mal, A. Chattopadhyay, A. Yang, R. Gadh, Electric vehicle smart charging and vehicle-to-grid operation. Int. J. Parallel, Emergent Distrib. Syst. **28**(3), 249–265 (2013). https://doi.org/10.1080/17445760.2012.663757
4. A. Vandermeulen, B. van der Heijde, L. Helsen, Controlling district heating and cooling networks to unlock flexibility: a review. Energy **151**, 103–115 (2018). ISSN: 03605442. https://doi.org/10.1016/j.energy.2018.03.034
5. A. Lever, H. Evans, M. Ravishankar, et al., *Flexibility in Great Britain. Imperial College London Consultants and The Carbon Trust* (2021), p. 201
6. N. Good, E.A. Martínez Cesena, P. Mancarella, Ten questions concerning smart districts (2017). https://doi.org/10.1016/j.buildenv.2017.03.037
7. L. Gkatzikis, I. Koutsopoulos, T. Salonidis, The role of aggregators in smart grid demand response markets. IEEE J. Select. Areas Commun. **31**(7), 1247–1257 (2013). ISSN: 07338716. https://doi.org/10.1109/JSAC.2013.130708
8. E. Dall'Anese, P. Mancarella, A. Monti, Unlocking flexibility: integrated optimization and control of multienergy systems. IEEE Power Energy Mag. **15**(1), 43–52 (2017). ISSN: 15407977. https://doi.org/10.1109/MPE.2016.2625218

9. A. Parisio, E. Rikos, L. Glielmo, A model predictive control approach to microgrid operation optimization. IEEE Trans. Control Syst. Technol. **22**(5), 1813–1827 (2014). ISSN: 10636536. https://doi.org/10.1109/TCST.2013.2295737
10. A. Moser, D. Muschick, M. Gölles, et al., A MILP-based modular energy management system for urban multi-energy systems: Performance and sensitivity analysis. Appl. Energy (2020). ISSN: 03062619. https://doi.org/10.1016/j.apenergy.2019.114342
11. M. Taylor, Optimal Coordination of District-Scale Multi-Energy Systems using Multi-Agent Control Architecture. PhD Thesis, University of Manchester (2022). [Online]. Available: https://research.manchester.ac.uk/files/261207751/FULL_TEXT.PDF
12. M. Taylor, O. Marjanovic, A. Parisio, Decentralized supervisory control of networked multienergy buildings. IEEE Transactions on Control Systems Technology **32**(3), 945–959 (May2024). https://doi.org/10.1109/TCST.2023.3346829
13. A. Parisio, C. Del Vecchio, A. Vaccaro, A robust optimization approach to energy hub management. Int. J. Electr. Power Energy Syst. **42**(1), 98–104 (2012). ISSN: 01420615. https://doi.org/10.1016/j.ijepes.2012.03.015. [Online]. Available: http://linkinghub.elsevier.com/retrieve/pii/S0142061512000701
14. H. Shi, N. Blaauwbroek, P.H. Nguyen, R.I. Kamphuis, Energy management in multicommodity smart energy systems with a greedy approach. Appl. Energy **167**, 385–396 (2016). ISSN: 0306-2619. https://doi.org/10.1016/J.APENERGY.2015.11.101
15. A. Parisio, C. Wiezorek, T. Kyntäjä, J. Elo, K. Strunz, K.H. Johansson, Cooperative MPC-based energy management for networked microgrids. IEEE Trans. Smart Grid **8**(6), 3066–3074 (2017). ISSN: 19493053. https://doi.org/10.1109/TSG.2017.2726941
16. W. Gu, Z. Wang, Z. Wu, Z. Luo, Y. Tang, J. Wang, An online optimal dispatch schedule for CCHP microgrids based on model predictive control. IEEE Trans. Smart Grid **8**(5), 2332–2342 (2017). ISSN: 19493053. https://doi.org/10.1109/TSG.2016.2523504
17. A. Anvari-Moghaddam, A. Rahimi-Kian, M.S. Mirian, J.M. Guerrero, A multi-agent based energy management solution for integrated buildings and microgrid system. Appl. Energy **203**, 41–56 (2017). ISSN: 0306-2619. https://doi.org/10.1016/J.APENERGY.2017.06.007
18. E.A. Cesena, P. Mancarella, Energy systems integration in smart districts: robust optimisation of multi-energy flows in integrated electricity, heat and gas networks. IEEE Trans. Smart Grid (2018). ISSN: 19493053. https://doi.org/10.1109/TSG.2018.2828146
19. J. Iria, F. Soares, Real-time provision of multiple electricity market products by an aggregator of prosumers. Appl. Energy **255**, 113–792 (2019). ISSN: 0306-2619. https://doi.org/10.1016/J.APENERGY.2019.113792
20. D. Ioli, A. Falsone, A.V. Papadopoulos, M. Prandini, A compositional modeling framework for the optimal energy management of a district network. J. Process Control. **74**, 160–176 (2019). ISSN: 0959-1524. https://doi.org/10.1016/J.JPROCONT.2017.10.005
21. S. Long, O. Marjanovic, A. Parisio, Generalised control-oriented modelling framework for multi-energy systems. Appl. Energy **235**, 320–331 (2019). ISSN: 03062619. https://doi.org/10.1016/j.apenergy.2018.10.074
22. T. Pippia, J. Sijs, B. De Schutter, A single-level rule-based model predictive control approach for energy management of grid-connected microgrids. IEEE Trans. Control Syst. Technol. 1–13 (2019). ISSN: 1063–6536. https://doi.org/10.1109/tcst.2019.2945023
23. N. Good, P. Mancarella, Flexibility in multi-energy communities with electrical and thermal storage: a stochastic, robust approach for multi-service demand response. IEEE Trans. Smart Grid **10**(1), 503–513 (2019). https://doi.org/10.1109/TSG.2017.2745559
24. R. Jing, M.N. Xie, F.X. Wang, L.X. Chen, Fair P2P energy trading between residential and commercial multi-energy systems enabling integrated demand-side management. Appl. Energy **262**, 114–551 (2020). ISSN: 0306-2619. https://doi.org/10.1016/J.APENERGY.2020.114551
25. A. Coelho, J. Iria, F. Soares, Network-secure bidding optimization of aggregators of multi-energy systems in electricity, gas, and carbon markets. Appl. Energy **301**, 117–460 (2021). ISSN: 0306-2619. https://doi.org/10.1016/J.APENERGY.2021.117460
26. G. Yuan, Y. Gao, B. Ye, Optimal dispatching strategy and real-time pricing for multi-regional integrated energy systems based on demand response. Renew. Energy. **179**, 1424–1446 (2021). ISSN: 0960-1481. https://doi.org/10.1016/J.RENENE.2021.07.036

27. H. Zakernezhad, M.S. Nazar, M. Shafie-khah, J.P. Catalão, Optimal resilient operation of multi-carrier energy systems in electricity markets considering distributed energy resource aggregators. Appl. Energy **299**, 117–271 (2021). ISSN: 0306-2619. https://doi.org/10.1016/J.APENERGY.2021.117271
28. L.M.P. Ghilardi, A.F. Castelli, L. Moretti, M. Morini, E. Martelli, Co-optimization of multi-energy system operation, district heating/cooling network and thermal comfort management for buildings. Appl. Energy **302**, 117–480 (2021). ISSN: 0306-2619. https://doi.org/10.1016/J.APENERGY.2021.117480
29. M. Wirtz, L. Neumaier, P. Remmen, D. Müller, Temperature control in 5th generation district heating and cooling networks : an MILP-based operation optimization. Appl. Energy **288**, 116–608 (2021). ISSN: 0306-2619. https://doi.org/10.1016/j.apenergy.2021.116608
30. A. Bemporad, M. Morari, Control of systems integrating logic, dynamics, and constraints. Automatica **35**(3), 407–427 (1999). ISSN: 00051098. https://doi.org/10.1016/S0005-1098(98)00178-2
31. HYSDEL—Hybrid System DEscription Language. [Online]. Available: http://people.ee.ethz.ch/%7B~%7Dcohysys/hysdel/ (visited on 10/22/2021)
32. S. Frank, S. Rebennack, IIE Transactions an introduction to optimal power flow: Theory, formulation, and examples an introduction to optimal power flow: theory, formulation, and examples (2016). ISSN: 1545-8830. https://doi.org/10.1080/0740817X.2016.1189626
33. A. Osiadacz, *Simulation and Analysis of Gas Networks* (E. & F.N, Spon Ltd, London, 1987). 0-419-12480-2
34. M. Pirouti, *Modelling and Analysis of a District Heating Network* (2013) p. 121. [Online]. Available: http://orca.cf.ac.uk/45201/1/Marouf.Pirouti%7B%5C_%7DPhD%20Thesis.pdf%20h
35. L. Gan, N. Li, U. Topcu, S.H. Low, Exact convex relaxation of optimal power flow in radial networks. IEEE Trans. Autom. Control **60**(1), 72–87 (2015). ISSN: 00189286. https://doi.org/10.1109/TAC.2014.2332712
36. M.K. Singh, V. Kekatos, Optimal scheduling of water distribution systems. IEEE Trans. Control Netw. Syst. **7**(2), 711–723 (2020). ISSN: 2325-5870. https://doi.org/10.1109/TCNS.2019.2939651
37. G.T. Ayele, P. Haurant, B. Laumert, B. Lacarri'ere, An extended energy hub approach for load flow analysis of highly coupled district energy networks: illustration with electricity and heating. Appl. Energy **212**, 850–867 (2018). ISSN: 03062619. https://doi.org/10.1016/j.apenergy.2017.12.090
38. S. Huang, W. Tang, Q. Wu, C. Li, Network constrained economic dispatch of integrated heat and electricity systems through mixed integer conic programming. Energy **179**, 464–474 (2019). ISSN: 03605442. https://doi.org/10.1016/j.energy.2019.05.041
39. D. Xu, Q. Wu, B. Zhou, C. Li, L. Bai, S. Huang, Distributed multi-energy operation of coupled electricity, heating, and natural gas networks. IEEE Trans. Sustain. Energy **11**(4), 2457–2469 (2020). ISSN: 19493037. https://doi.org/10.1109/TSTE.2019.2961432
40. Y. Zhou, M. Shahidehpour, Z. Wei, Z. Li, G. Sun, S. Chen, Distributionally robust unit commitment in coordinated electricity and district heating networks. IEEE Trans. Power Syst. **35**(3), 2155–2166 (2020). ISSN: 15580679. https://doi.org/10.1109/TPWRS.2019.2950987
41. Y. Cao, W. Wei, L. Wu, S. Mei, M. Shahidehpour, Z. Li, Decentralized operation of interdependent power distribution network and district heating network: a market-driven approach. IEEE Trans. Smart Grid 1 (2018). ISSN: 19493053. https://doi.org/10.1109/TSG.2018.2880909
42. S. Yao, W. Gu, S. Lu, et al., Dynamic optimal energy flow in the heat and electricity integrated energy system. IEEE Trans. Sustain. Energy **12**(1), 179–190 (2021). ISSN: 19493037. https://doi.org/10.1109/TSTE.2020.2988682
43. K. Lygnerud, S. Werner, Annex TS2 implementation of low-temperature district heating systems. Final Report. IEA-DHC, Tech. Rep. (2021)
44. H. Lund, S. Werner, R. Wiltshire, et al., 4th Generation District Heating (4GDH). Integrating smart thermal grids into future sustainable energy systems. Energy **68**, 1–11 (2014). ISSN: 03605442. https://doi.org/10.1016/j.energy.2014.02.089

45. S. Buffa, M. Cozzini, M. D'Antoni, M. Baratieri, R. Fedrizzi, 5th generation district heating and cooling systems: a review of existing cases in Europe. Renew. Sustain. Energy Rev. **104**, 504–522 (2019). ISSN: 1364-0321. https://doi.org/10.1016/J.RSER.2018.12.059
46. M. Taylor, S. Long, O. Marjanovic, A. Parisio, Model predictive control of smart districts with fifth generation heating and cooling networks. IEEE Trans. Energy Convers. 1–11 (2021). ISSN: 0885-8969. https://doi.org/10.1109/tec.2021.3082405
47. E.F. Camacho, C. Bordons, Model predictive control (2007), p. 405. ISBN: 9781852336943. https://doi.org/10.1007/978-0-85729-398-5. arXiv: arXiv:1011.1669v3
48. L. Wang, Model predictive control: design and implementation using MATLAB (2009), pp. 25–26. ISBN: 9781848823303. https://doi.org/10.1109/acc.2009.5159781. arXiv: arXiv:1011.1669v3
49. D.E. Seborg, T.F. Edgar, D.A. Mellichamp, F.J. Doyle, *Process Dynamics and Control*, 3rd. Wiley (2010)
50. J.M. Maciejowski, *Predictive Control: With Constraints* (2002), p. 331. ISBN: 0201398230. [Online]. Available: http://books.google.com/books?id=HV_Y58c7KiwC&pgis=1
51. J. Reynolds, M.W. Ahmad, Y. Rezgui, J.L. Hippolyte, Operational supply and demand optimisation of a multi-vector district energy system using artificial neural networks and a genetic algorithm. Appl. Energy **235**, 699–713 (2019). ISSN: 0306-2619. https://doi.org/10.1016/J.APENERGY.2018.11.001
52. P. Velarde, L. Valverde, J.M. Maestre, C. Ocampo-Martinez, C. Bordons, On the comparison of stochastic model predictive control strategies applied to a hydrogen-based microgrid. J. Power Sources **343**, 161–173 (2017). ISSN: 03787753. https://doi.org/10.1016/j.jpowsour.2017.01.015
53. M. Petrollese, L. Valverde, D. Cocco, G. Cau, J. Guerra, Real-time integration of optimal generation scheduling with MPC for the energy management of a renewable hydrogen-based microgrid. Appl. Energy **166**, 96–106 (2016). ISSN: 03062619. https://doi.org/10.1016/j.apenergy.2016.01.014
54. Y. Liu, N. Yu, W. Wang, et al., Coordinating the operations of smart buildings in smart grids. Appl. Energy. **228**, 2510–2525 (2018). ISSN: 03062619. https://doi.org/10.1016/j.apenergy.2018.07.089
55. R. Vujanic, P. Mohajerin Esfahani, P. J. Goulart, S. Mariéthoz, M. Morari, A decomposition method for large scale MILPs, with performance guarantees and a power system application. Automatica **67**, 144–156 (2016). ISSN: 00051098 https://doi.org/10.1016/.automatica.2016.01.006. arXiv:1411.1973
56. A. Falsone, K. Margellos, M. Prandini, A decentralized approach to multi-agent MILPs: finite-time feasibility and performance guarantees. Automatica **103**, 141–150 (2019). ISSN: 00051098. https://doi.org/10.1016/j.automatica.2019.01.009. arXiv: 1706.08788
57. T. Sommer, M. Sulzer, M. Wetter, A. Sotnikov, S. Mennel, C. Stettler, The reservoir network: A new network topology for district heating and cooling. Energy **199**, 117–418 (2020). ISSN: 03605442. https://doi.org/10.1016/j.energy.2020.117418
58. J. Skaf, S.P. Boyd, Design of affine controllers via convex optimization. IEEE Trans. Automatic Control **55**(11), 2476–2487 (2010). ISSN: 00189286. https://doi.org/10.1109/TAC.2010.2046053
59. R.D. Zimmerman, C.E. Murillo-Sanchez, MATPOWER manual, pp. 0–249 (2020). [Online]. Available: https://matpower.org/docs/MATPOWER-manul.pdf
60. A. Prasanna, V. Dorer, N. Vetterli, Optimisation of a district energy system with a low temperature network. Energy **137**, 632–648 (2017). ISSN: 03605442. https://doi.org/10.1016/j.energy.2017.03.137
61. U.S. Department of Energy, Commercial Reference Buildings (2016)
62. Octopus Energy Ltd, Agile Octopus rates. [Online]. Available: https://s3-eu-west-1.amazonaws.com/octoenergy-production-statics/data/agile-rates/agile_rates.2018-12-20.xlsx (visited on 02/04/2020)
63. Gurobi Optimization LLC, Gurobi Optimizer Reference Manual (2021). [Online]. Available: https://www.gurobi.com

64. J. Löfberg, Yalmip : a toolbox for modeling and optimization in matlab, in *In Proceedings of the CACSD Conference* (Taipei, Taiwan, 2004)
65. The MathWorks Inc., *MATLAB & Simulink*. [Online]. Available: https://uk.mathworks.com/products/matlab.html

SPRINGER NATURE

GPSR Compliance

The European Union's (EU) General Product Safety Regulation (GPSR) is a set of rules that requires consumer products to be safe and our obligations to ensure this.

If you have any concerns about our products, you can contact us on ProductSafety@springernature.com

In case Publisher is established outside the EU, the EU authorized representative is:

Springer Nature Customer Service Center GmbH
Europaplatz 3
69115 Heidelberg, Germany

The manufacturer's authorised representative in the EU is Springer Nature Customer Service Centre GmbH, Europaplatz 3, 69115 Heidelberg, Germany. If you have any concerns regarding our products, please contact ProductSafety@springernature.com

Printed and bound by CPI Group (UK) Ltd, Croydon, CR0 4YY

30/09/2026

02225253-0002